Click
with Your
Chick

CLICK WITH YOUR CHICK

CompanionHouse Books™ is an imprint of Fox Chapel Publishers International Ltd.

Project Team
Vice President-Content: Christopher Reggio
Editor: Amy Deputato
Copy Editor: Jeremy Hauck
Design: Mary Ann Kahn
Index: Elizabeth Walker

ISBN 978-1-62008-344-4

Library of Congress Cataloging-in-Publication Data

Names: Keyes, Giene, author.
Title: Click with your chick : a complete chicken training course using the
  clicker / Giene Keyes.
Description: Mount Joy, PA : Fox Chapel Publishing, [2019] | Includes index.
  | Identifiers: LCCN 2019018173 (print) | LCCN 2019020457 (ebook) | ISBN
  9781620083451 () | ISBN 9781620083444 (softcover)
Subjects: LCSH: Chickens--Training. | Clicker training (Animal training)
Classification: LCC SF487 (ebook) | LCC SF487 .K5615 2019 (print) | DDC
  636.5/1--dc23
LC record available at https://lccn.loc.gov/2019018173

This book has been published with the intent to provide accurate and authoritative information in regard to the subject matter within. While every precaution has been taken in the preparation of this book, the author and publisher expressly disclaim any responsibility for any errors, omissions, or adverse effects arising from the use or application of the information contained herein. The techniques and suggestions are used at the reader's discretion and are not to be considered a substitute for veterinary care. If you suspect a medical problem, consult your veterinarian.

Fox Chapel Publishing
903 Square Street
Mount Joy, PA 17552

Fox Chapel Publishers International Ltd.
7 Danefield Road, Selsey (Chichester)
West Sussex PO20 9DA, U.K.

www.facebook.com/companionhousebooks

We are always looking for talented authors. To submit an idea, please send a brief inquiry to acquisitions@foxchapelpublishing.com

Printed and bound in China
22 21 20 19   2 4 6 8 10 9 7 5 3 1

Click
with Your
Chick

A Complete Chicken Training
Course Using the Clicker

Giene Keyes

Contents

Acknowledgments

There are many people I'd like to acknowledge for their contributions to my work on this book, for their support in building my confidence to write it, and for their insights on chickens.

I would like to thank Heather Lockhart for her chicken expertise. Heather may not admit it, but she pretty much knows just about everything there is to know about chickens. She's my go-to chicken expert, and she's always willing to answer questions and spend as much time with me as I need. She's a pretty awesome lady.

A big thanks to Susan Troller for inviting me to do chicken-training seminars and (half-jokingly) explaining that if chickens were large enough, they'd eat us—which I truly believe! When Susan owned Cluck the Chicken Store, she invited me to to show off how clever chickens can be and to recruit as many people to fall in love with chickens as I could.

Thank you to my dear friend Stephanie for her love of anything with feathers, fur, toes, and snouts. The passion in her voice when she talks about the animals she loves is enough to convince anyone to love them just as much.

I'd like to send an enormous thank you to all of the sanctuaries and rescues that are trying to educate the public and spread the word about the plight of the poor chicken, a magnificent and intelligent creature worthy of respect and consideration.

I'd especially like to thank my husband and my kids. My husband let me get my first chicken, and my second, and my fortieth. He lets me do crazy things like go to poultry swaps (and then forbids me from ever going again when he sees what I come home with). He lets me keep a chicken in the house because he knows it makes me happy. Actually, he doesn't "let" me, he supports me. And my three kids, who probably roll their eyes whenever they hear me go on a twenty-minute-long speech when someone asks, "Why would you train a chicken?" My daughter is my amazing photographer, and she is very patient with me when I want to get it "just right." My oldest son has an aura about him that animals flock to. They all feel so comfortable around him. No matter the species, they all just seem to want to be with him, doing whatever he's doing. And my youngest child is my little chicken whisperer, my partner in crime. I think he loves chickens more than anyone I've met, and he allows me to take a moment to just think of the joy that these creatures can bring.

Introduction

The idea for this book is one that has been nesting for a while. When I first got chickens, I didn't realize that I wanted to write a book on how to train and bond with them, but I've enjoyed having chickens more than I could have imagined. Through the years, I've had chickens come, and I've had chickens go. I've lost some of my favorite chickens. Two passed suddenly, and I've lost some to predators. To be honest, I didn't start training chickens because I loved them; I initially started training them because I wanted to become a better dog trainer. I truly didn't realize how much fun I'd have training my little dinosaurs! Even more so, I had fun just simply being around my chickens, observing them, spending time with them, and getting to know them. The more people with whom I talk about chickens, the more they are interested in them, too—not just training them, but really understanding them and bonding with them.

As I progressed in my training, I started holding seminars and chicken-training workshops, and I began to realize that each one of my hens had her own little personality. Even chickens within the same breed have distinct personalities. You may think to yourself, "Duh! Of course they each have a different personality!" But to many people, it may sound odd. Many people look at a group of chickens and honestly don't see past a group of chickens. It's not because they don't really care about the chickens; in fact, they may really love animals. However, I think that, as a society, we group animals together as one and don't really consider that they are individual beings. I'll admit that when they all look the same, it's hard to differentiate. But once you get to know them—well, you know! They certainly are all different little creatures with their own likes and dislikes. They even form close bonds with their friends.

I got my first chickens in 2012. When I write it, it doesn't seem like that long ago, yet it feels as if I've had chickens forever. I grew up in the city but have always been a country girl at heart, so maybe that's why. Whatever the reason, I'm hooked! It seems as if more and more people I know are joining me in chicken ownership; there are backyard, urban, 4-H, and rural chicken lovers. My passion for

animal behavior is what led me to chickens. I have studied dog (and human) behavior for more than twenty years. During that time, I owned a dog daycare and training company. I have trained thousands—yes, thousands!—of dogs and their people. Dog training is a pretty cool profession because there is always something new to learn.

Around 2008, I heard about "chicken camps." Bob Bailey and Terry Ryan both have held these chicken camps around the world (yes, it's a thing).

Birdie, one of our very first chickens.

Dog trainers love to go to chicken camps because the camps improve their ability to train dogs. Chickens are crazy fast, and you have to keep their attention or else they will simply walk away (unlike a dog, over whom you have some sort of control because he is leashed). Working with chickens, you learn to improve your timing, observation skills, and much more. I had always wanted to go to a chicken camp but didn't have the time or the resources to do so. So, a few years after we moved to the country, I decided to get a flock of my own. As a child, I was always asking my parents for this animal or that animal, and my mom told me that I could have as many animals as I wanted when I was grown up—time to get some chickens!

A couple days before Easter (cliché, I know...), I went to a friend's house and picked out five little adorable chicks that were no larger than the palm of my hand. At the time, I had no idea how much joy these little creatures would bring me and my family. They have introduced me to a whole new world, including amazing animal behaviors, great new friends, and a renewed appreciation for an animal that is very basic yet continues to amaze me with its intelligence. In fact, my friend Susan Troller has a wonderful analogy to explain the intelligence of chickens. She says that they have little computer chips for brains, and those little chips hold millions of years of experience and information!

When I was younger, I always craved the knowledge to understand how to train animals. We had a family dog who certainly learned from her mistakes. She would get into the garbage can and then be put in the basement for a long time-out. Maybe hunting and show dogs had crates

Me with my friend Susan Troller, who once owned a chicken-supply store.

at that time, but no one knew about using them with family dogs. I would have loved to read a book on dog training or to have taken my dog to a training class, but we didn't have anything like that around us. I remember going to the library to find books on dog training, but the only ones I could find were on how to train "bird dogs." I wanted to know not only how to train my dog but also how to train her in a non-forceful manner.

When I was nine years old, I got my first horse. I loved that horse more than life itself. I started taking riding lessons at the stable on the outskirts of town. I remember one day when my horse was acting up. The instructor got on her, beat her with the whip, and made her gallop around the ring about twenty times. She was sweaty and out of breath, and I felt like all of the blood had been drained from my body. Again, I craved the knowledge on how to train her and how to handle that situation better.

When I was twenty-five, I got my very own puppy. I was so excited to train this chubby little black Labrador that the first thing I did was sign up for puppy class. I was such a training-class geek! I had my leash, my treats, and a mile-wide smile. When I arrived at class, an instructor took off my puppy's collar and slapped a choke chain on him. I remember feeling like that wasn't right, but the instructors were the professionals, after all, so I did what they told me to do. Yank up for sit; yank down for down. We progressed through the levels of the classes. My dog learned, but he certainly was not an operant (see definitions starting on page 72) dog, nor was he happy to be in class.

I remember the straw that broke the camel's back: We had gotten to the intermediate level and were starting to learn how to heel off leash. The instructor came over to me and told me to drop my leash and put my thumb and index finger on my dog's ear. She said, "If he gets too far away from you, you pinch his ear as hard as you can." That was it! I couldn't do this type of training anymore. I wanted my dog to be happy and willing and to want to do things with me. I knew there must be another way. From that moment on, even though I didn't discover clicker training until years later, I used positive-reinforcement training—with horses, pigs, cats, chickens, husbands, kids, you name it!

I do have to admit, training chickens has made me a bit humble. I can see how much of a slow human being I am sometimes. I sometimes think, "I just missed the opportunity to reward that behavior!" But chickens are pretty forgiving. They will still learn behaviors even though you may be the one who has to catch up to them. After a while, you and your chicken will start to work together as a team, learning from one another. It's a pretty cool thing.

Training your chicken is only partly about observing behaviors and honing in on your timing skills. The tricks are super cool. The amazement you'll feel when your chicken "gets it" is awesome. But the truly wonderful part is that you'll find yourself bonding with your chicken, and your chicken will be bonding with you. She'll be engaged with you, and she'll want to spend more time with you. My guess is that you'll want to spend more time with her, too!

Just to warn you: this is not a book about how to keep chickens. It's not about how to care for your chicken (other than giving her love and encouragement, which is certainly part of the book!). It's not about how to build a coop. There are shelves full of books about chickens and how to care for them. This is a book about training your chicken and how to really connect and engage with your bird—something you'll both find very rewarding!

I'm glad you're here with me to learn about training your chicken. It may sound silly to some people, but once you start doing it, I think you'll be in awe of these little creatures just as I have become.

One of the many coops my talented husband has built for our chickens.

Chapter 1

How
Chickens Learn

We breed Portuguese Water Dogs. They are an amazing breed, and I always love to show them off and talk about how much we love them. But when people come to our home to meet our dogs, it's almost inevitable that we end up in the backyard with the pigs and the chickens. A family with young children recently came to visit, so I took a bag of leftover food from the refrigerator, and we went outside to feed the pigs and the chickens. As we were feeding them, I didn't even realize that I kept offering little bits of information about the chickens. I'd say things like, "See how the rooster is making that cute little sound? He's actually calling the hens right now to tell them that there is yummy food over there." The woman commented by saying, "Wow, you really know a lot about chickens!" I had to laugh at myself because I never quite realize when I have been going on and on about the

Me with Portuguese Water Dogs Axl, Liberty, and Tallulah.

Some of our flock with our pigs, munching on a seed block together.

chickens. I smiled and explained to her that I find them so interesting. I told her that the more time I spend with my chickens, the more I'm amazed at what cool creatures they are. There are so many amazing things to learn about chickens!

Chickens are highly intelligent, social beings. We are lucky to have so much scientific research about them at our fingertips. Think about how much we know about chickens today and how much more we will know twenty years from now. Here are some of the interesting things that science tells us about chickens.

They have more than thirty different communication sounds. I recognize sounds of baby chicks calling for their mom. I also recognize sounds of a hen announcing to the world that she is about to lay an egg. There are sounds for mothers calling for chicks, and sounds for roosters crowing (morning, noon, and night!). I hear chickens making cooing sounds to each other when they're standing around or calling to each other when they have found yummy treats on the ground. I hear sounds of boredom and sounds of excitement. I have heard my rooster call to the girls to try to get their attention (while the girls completely ignore him). I've also heard him yell at them that a hawk is overhead and then watch them all scurry into the barn. I've heard my chickens communicate when it's evening and time to perch. Gosh, I can think of around fifteen different sounds just from general observation!

They have a social hierarchy. It's basically a means of keeping order in the flock. Usually, there is an alpha hen and an alpha rooster. Right now, I have two roosters (father and son,

The one and only Sprinkles!

Baby Checkers, vying to be "top hen."

and sometimes they act like it!). I have to admit, my alpha rooster seems pretty tolerant. He has never beaten up on any of the hens, and his son seems to pretty much ignore him. Every once in a while, the alpha will walk over to his son and just give him a reminder of who's boss with a fairly gentle peck on the head (hence the phrase "pecking order"). I also have my little bantam house-chicken, Sprinkles, indoors with me. She must be (or must want to be) an alpha hen because she has taken on the responsibility of crowing! I'm not sure if I should call her a "hooster" or a "rhen."

Sometimes I'll pluck a chicken from the barn to bring her in the house with me for training. Knowing that chickens have a social hierarchy, the pecking order changes with one of them gone, and it then has to be established again when the chicken returns. If the chicken that leaves the group is the alpha chicken, the second in line has to be ready physically and mentally to take on that role.

Chickens are said to have emotional intelligence. Emotional intelligence is the ability to identify and manage one's emotions as well as the emotions of others. The three components of emotional intelligence are emotional awareness; the ability to harness emotions and apply them to tasks, such as thinking and problem solving; and the ability to manage emotions, including regulating one's own emotions and cheering up or calming down others.

It sounds a bit profound, doesn't it? Emotional intelligence in a chicken? Trying to explain this to non-chicken people might take a little convincing, but I can tell you that I've seen emotional intelligence in chickens firsthand, probably on a daily basis. Running for cover when your friend tells you that a predator is coming? I'd say that's a situation in which you'd be emotional and have to think rationally. Finding a delicious bug, becoming excited, and calling your girlfriend over so she can have it instead of you? Could that be categorized as cheering someone up? Well, I do know that my chickens cheer me up all the time. And then, with certain other emotions, I think my chickens are more like "whatever."

Chickens have problem-solving skills. If you think of the high intelligence level of birds in general, their problem-solving skills should not surprise you. A magpie can recognize its image in a mirror. African grey parrots can count, categorize objects by color and shape, and learn to understand human words (not just repeat them, but actually understand them). There are cute videos of cockatoos dancing and keeping the beat to music. Crows and ravens are clever problem-solvers and expert toolmakers. The same is true for chickens, and they demonstrate these skills in their daily activities, such as figuring out the pecking order and how to live peacefully among a flock.

Here's an example of problem solving: I used to have a chicken that would peck at my feet so I would feed her. I would go into the barn, she'd run over and peck at my feet, and I would, in turn, pour food in her bowl. This was her little routine, and she was pretty proud of the fact that she knew how to make me do something. Well, summer came, and I started wearing flip-flops into the barn. A chicken's pecks can be hard on bare toes, so I decided that I didn't want her to do that behavior anymore. If I would have reacted in any way to her pecking at my feet (shooing her away, walking around, or something similar), she would have continued to peck at my feet because she would be getting at least some type of response. So, for a few days, each time I went into the barn, I just stood there. The first day, she pecked at my feet for about ten minutes. She became very frustrated and even tried scratching at my feet. When she decided to stop pecking at my feet and just stood near me, I walked to her food container and fed her. The second day was the same story, but it only took about a minute for her to stop pecking at my feet. By the third day, she was no longer pecking at my feet. She had figured out what to do to get her food.

So, if birds in general are smart enough to understand humans, make tools, and keep a beat, then why do most people think of chickens as "bird brains?" I guess chickens are just egg-laying robots that don't have feelings or the brain capacity to think past laying and pecking,

Each chicken in the flock has his or her own personality, likes and dislikes, and feelings.

right? It might just be a lack of education. Looking at a group of animals all together, it can be hard for some people to really understand that each one of the animals in that herd, pack, or flock has its own personality, feelings, emotions, and friendships. Likewise, considering a group of chickens, all the same size and color, in a barn, I can see how it may be hard for some people to realize the birds' intelligence and individuality. I'm sure most people's first thought isn't, "Wow, those birds seem really smart!"

One of the most difficult and rewarding jobs I've had is when I owned a dog daycare. Part of the reason it was difficult is because when I started it in 2004, most people didn't know what a dog daycare was or why you'd want to take your dog there. One of my marketing strategies was to let my friends and their friends bring their dogs to daycare for free for the first two months. Because not many people in town knew about dog daycare, my "master plan" was to have potential clients enter an active building and see dogs running around, playing, and having a blast! My friends who brought their dogs enjoyed it so much (or, should I say, the dogs enjoyed it so much) that after the first couple of months for free, they became paying clients.

I mention my dog daycare because there would be times when we would have maybe twenty black Labradors or fifteen Golden Retrievers, all in the same room. Our daycare was set up so that owners and potential clients could walk into the room and be able to see all of the dogs interacting. I can't tell you how many times people would ask, "How do you tell all of the dogs apart?" We'd respond with what any parent or babysitter would say: "When you get to know their personalities, you can see the differences in them quite easily." It's not always just a matter of one having a longer snout or one having a curlier tail. It's really their personalities that set them apart from each other.

I'm certain that chickens develop lifelong friendships, just as humans, dogs, elephants, and many other species do. Geese mate for life. Elephants have been observed remembering friends they hadn't seen in more than thirty years. It's said that barn owls form emotional connections to their partners, and they even cuddle just to cuddle (not only as a breeding activity).

Can you imagine that the same would be true—maybe even stronger—for chickens? I certainly can. Probably a couple of times a year, I'll introduce a new chicken or group of chickens to my flock. In my experience, whenever I introduce a new chicken or group to my flock, the new chickens tend to stay to themselves. They are not truly integrated into the established flock. The new chickens create their own "mini-flock." Basically, they just stick together. There have also been instances in which one chicken from a bonded pair gets killed by a predator (yes, it's one of the unfortunate realities of keeping chickens), and that single chicken never really gains another friend after his or her buddy passes away. It's actually very sad. I think chickens have to grow up together in order to truly bond with/imprint on each other. I'm sure there are stories of chickens becoming pals later in life, but I just haven't seen it happen with my flock.

When training chickens, nurture plays a large role. Your chicken has to feel safe around you, and sometimes you have to teach him or her how to learn. It sounds silly, I know. As Sophia Yin states in a blog post (*https://drsophiayin.com/blog*), "... training of simple tasks just involves a few things—a hungry animal in a comfortable environment and a trainer with good timing."

I believe that chickens, just like any creature, learn in a number of different ways. It's part nature and part nurture. The nature part explains why different breeds of chickens act differently. For example, many bantam breeds are "flighty" and reactive. They may not be the best to train because they'll be rather jumpy and quick to try to escape. Having said that, my little Sprinkles is a bantam chicken, and she chooses to stay near me when she's out of her enclosure. She's

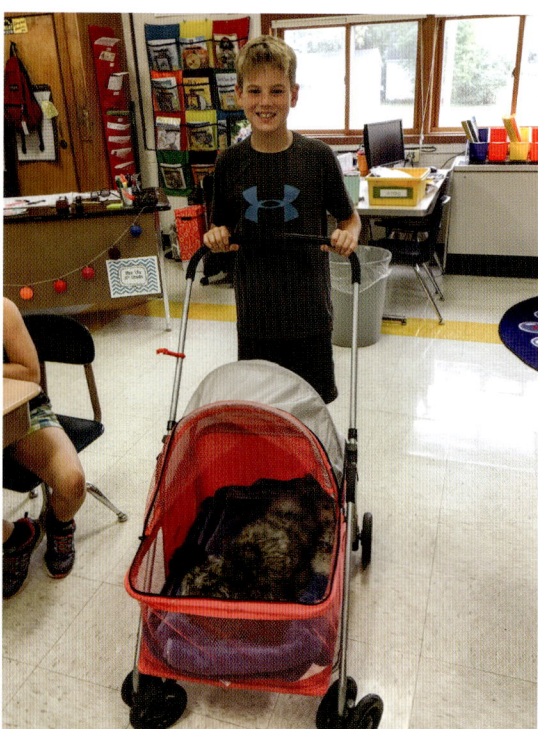

Silkies are wonderful chickens for visiting schools.

also extremely smart. She's a "singleton" chicken, meaning she was the only egg that hatched. Fortunately for me, she is probably more bonded with me because of it. If she had a flock of her own, you could argue that she may not have such a strong relationship with a human. Since we don't know for sure, I'm just going to say that she loves me as much as I love her! Then there is the adorable little Silkie, also a bantam breed. Silkies are fantastic for children because they are easy to pick up and are usually quite content to sit in your lap.

I believe that when you take a chicken out of the flock, you can really see his or her individual personality blossom. When chickens are in the flock, they are considering their pecking order and the regular habits of the day, such as searching for food, laying eggs, taking dust baths, and when to perch in the evening. But, when you take them out of that environment and provide them with food, structure, and an opportunity to experiment, you can really see your chickens' personalities.

Clicker Training

I like to use a clicker for my chicken training. Clickers have been popular among dog trainers since the early 1990s. They are also used to train dolphins, elephants, camels—you name it! If an animal can hear the click and accept a reward (usually a high-value treat), it is a creature that can be clicker-trained. There is scientific research that backs up the effectiveness of clicker training as well as countless studies on clicker training versus marker words. When using marker words,

The simplest type of clicker is made of metal encased in a plastic rectangle.

the trainer or owner says a word like "yes!" in place of the click. Most studies found marker words less effective than the clicker because the sound of a person's voice can vary—higher, lower, animated, excited, and so on—whereas a click is a constant.

When you're training species other than chickens—dogs, for example—you can use a variety of reward methods. Saying "good boy" or giving a scratch on the rear will usually make a tail wag, but try that with a chicken, and you'll most likely get a different response! As much as we would like to think that our chickens enjoy being around us, you'll need to give your chickens simply irresistible rewards when you're training them and want them to stay with you. A little scratch on the head probably won't be an effective reward for a chicken. Chickens seem to do best when you use the clicker to mark the behaviors you want. The "trick to the click" is to help your chickens form a strong positive association with the sound of the click. If you click (to mark) a chicken's behavior and then offer a reward that the chicken considers "blah," then the click won't truly mean anything to the chicken. But if your chicken simply can't resist cut-up strawberries, and she gets a piece of strawberry each time she hears the click, then you're going to create a highly motivated chicken.

The sound from the clicker is not the reward; it simply marks the behavior we want. It tells your chicken that she just did something right, and a reward is coming. The "click" marks the exact moment your chicken did the desired behavior. Your chicken hears the click and understands that what she was doing, at that very split second, is the behavior that is going to get her the food reward. In the beginning, you're going to want to click and reward a lot. And, in the beginning, the reward has to come immediately after the click.

SPRINKLES

Meet Sprinkles! We picked up Sprinkles from a woman I met on Facebook after I had posted to ask if anybody had a tiny chicken that needed a home. The woman replied and said that she had an adorable little girl who had hatched under a Frizzle at the county fair and was about a year old. She had been the only one to hatch, and she had gotten plenty of attention at the fair because, of course, nothing could be cuter than a day-old baby chick surrounded by a bunch of 4-H kids. The reason that the woman had decided to offer Sprinkles to a new home is because, first of all, Sprinkles had never really bonded with anybody, and, second, she was in a coop with a rooster and his mate, and the rooster was beating up on her a lot.

Being the animal lover that I am, and also being very used to driving long distances to pick up animals (from my days of fostering dogs and dog shows), I hopped in the car the next morning and drove an hour to go pick up this tiny chicken. My youngest son, who shares my love for animals, decided to come with me. He's my partner in crime because he likes to come with me to chicken swaps and to the zoo and to visit friends who are fostering dogs—he'll basically take any chance he can get to be with an animal.

The woman who had Sprinkles was very lovely and did a lot of fostering herself. As we pulled up in her driveway, six or seven dogs came running out to the car to greet us, and we were

Sprinkles loves looking for bugs!

Dot and Dusty.

thrilled to see each one of them. We got out of the car and went into her backyard, where she had three adorable coops, each with its own little flock of chickens. She opened up the coop door, and we met Sprinkles for the first time. She is half Old English Game Bantam and half Mille Fleur Belgian Bearded d'Uccle Bantam. She's also my first bantam, so she seems extra tiny to me.

My husband, being the accommodating, tolerant person that he is, doesn't mind that Sprinkles lives in the house with us. We built her an adorable little indoor coop because I wanted her to be easily accessible to do a lot of training with me. So now we have a cute little 1½-pound house chicken named Sprinkles, and I get to explain to my friends why it's so much fun to have a chicken in the house. They all pretty much think I'm a crazy animal lady, anyway!

As of the time of writing, Sprinkle laid eggs, and two of them hatched. We have little teeny twins! We named one Dot and one Dusty. I don't know how to tell if bantams are roosters when they're born, but little Dusty was a spitfire right out of the egg. He's also the louder of the two, and he looks up at you when you approach. So, my guess is that he's a little roo. They are both

When you're first teaching your chicken how to learn, you're going to be clicking and treating for just about any behavior your chicken offers that is something you like. Obviously, you're not going to click and treat your chicken for flying off of the training table (unless, of course, that is what you're training her to do). Actually, that might be a cool trick—you can go through your chicken's whole repertoire and then have her fly off of the table as a finale!

Jackpots

A jackpot basically tells the animal that what he or she just did was super awesome! Say you were working on sit with your dog. Each time you tried it, your dog almost—but not quite—did it. Then, maybe on the twentieth try, he finally got it! Yippee! That would be a perfect time for a jackpot. Take five or six treats in your hand but don't give them to your dog all at once. Give them one at a time, in super quick succession. Your dog (or chicken) will think, "This is the best thing ever!" It's kind of like when a human plays a slot machine. When you get the jackpot, you don't get a $100 bill; you get quarter after quarter after quarter—much more exciting, right?

To give your chicken a jackpot, hold the treat container still, right in front of her, so she doesn't have to take a step toward it, and let her peck a couple of times. Then start to remove the treat container, but give it right back to her and let her peck two or three more times. Repeat the process three or four times. It should really only take a few seconds from start to finish, but in your chicken's mind, she'll be thinking, *Wowza! This is awesome*! And she'll realize that the behavior she just did is something you really like because she gets a whole bunch of treats.

To offer your chicken a jackpot for a job well done, allow her to peck a few times at the treats.

Behavior Extinction

Something else you'll see during training is called behavior extinction. This is when an animal has learned a behavior, be it good or bad, that you no longer want the animal to do. For example, I want my chicken to peck at a dot on a paper plate, so I reward her for doing so. She learns this new trick fairly easily and now pecks at the dot each time it is presented to her. Now, I want to change the behavior so that, when she sees the dot on the plate, she walks around the plate. My little chicken is a clever girl, but she's not a mind reader! So, when presented with the dot on the plate, she is going to continue to peck at it. However, I am not going to reward her because I have changed the criteria. Most likely, she will continue to peck at the dot. She may peck harder, and she may peck faster. She may become frustrated and rub her beak on the dot. She may become even more frustrated (because she is not getting the click and treat) and scratch at the dot. Then, finally, at some point, she will not do anything when presented with the dot. That is what behavior extinction looks like. When you get to that point, you can start rewarding her for baby steps toward the new behavior of walking around the plate with the dot.

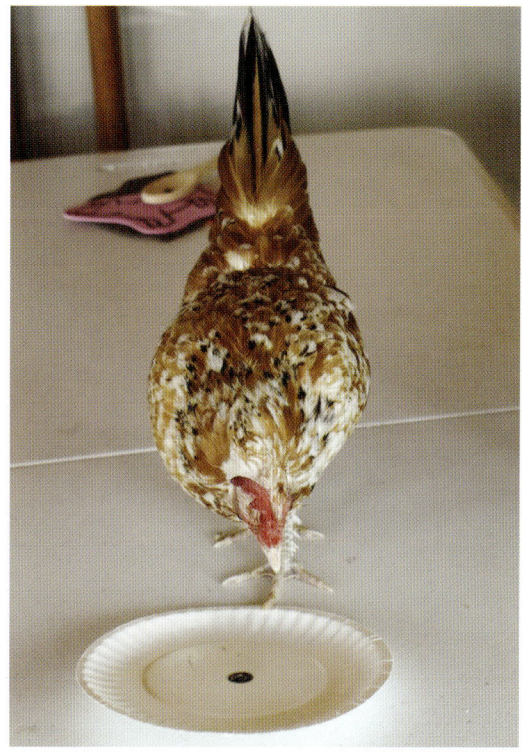

A behavior may get stronger before the chicken stops doing it; this is a principle of behavior extinction.

Operant Conditioning

In this book, we are going to talk about a few different methods of training our chickens, including operant conditioning, shaping, chaining, and target training. The one that you'll probably use the most is operant conditioning. If you have an opportunity to attend a seminar by Bob Bailey, I highly recommend it. He is one of the experts on operant conditioning.

Operant conditioning is a process in which animals (and husbands and children) learn to behave in a way that will get them a reward. For example, a dog barks at the door, and his owner lets him out. Animals also use operant conditioning to avoid punishments. For example, a horse learns to "give," or turn his head, when pressure is applied to the bit in his mouth in order to make the pressure stop. The animal becomes "operant." Remember that behavior is controlled by consequences. The dog learns that if he barks, the door will open. The horse learns that if he turns his head, the pressure in his mouth will stop.

REINFORCEMENT AND PUNISHMENT

During our chicken training, we are going to use positive reinforcement. It's unlikely that a chicken would learn anything from negative reinforcement or negative punishment—more likely, she would just shake a tail feather and walk away. Plus, we don't want to train our chickens (or any animals) through punishment. Studies show that chickens have good memories and will avoid situations that they remember as negative or dangerous. So, positive-reinforcement training it is! This term is used rather broadly, so I'm going to break it down to relate to our chicken training.

REINFORCEMENT

Trainers use reinforcement to help increase the probability that a specific behavior will occur in the future by delivering or removing a stimulus immediately after the behavior. More simply put, reinforcement, if done correctly, results in a behavior occurring again in the future. The reinforcement can be anything that the animal finds pleasing. My dogs find pieces of steak pleasing. They also enjoy attention and physical praise. My chickens also find pieces of meat pleasing, and they will work for mealworms and graham cracker crumbs as well. They don't find physical praise pleasing, so that would not be considered "positive reinforcement" for them.

Reinforcement can be positive or negative. Let's take a closer look at both types of reinforcement.

Positive reinforcement: Positive reinforcement works by presenting a motivating/reinforcing stimulus to the chicken after she displays the desired behavior, making the behavior more likely to happen in the future. For example, when you call "chick-chick," your chicken comes running because each time you say "chick-chick," you leave yummy treats for her. She is more likely to display this behavior in the future.

Negative reinforcement: Negative reinforcement occurs when a certain stimulus (usually an aversive stimulus) is removed after the animal displays a particular behavior. The likelihood of the behavior occurring again in the future is increased because either you've taken away the negative consequence or the animal can avoid the negative consequence. The example of the horse and the bit that I previously mentioned is an example of negative reinforcement. The horse learns to "give," or turn his head, to avoid the pressure on the bit in his mouth when the reins are pulled back. Another example of negative reinforcement would be my husband building a new chicken coop to avoid listening to me asking him each weekend (just kidding, of course!).

PUNISHMENT

When you hear about using punishment in training, you usually think of it as being something bad, but this is not always the case. Trainers use punishment to help decrease the probability that a specific behavior will occur in the future—in other words, a consequence immediately following a behavior that decreases the future frequency of that behavior. As with reinforcement, a stimulus can be added (positive punishment) or removed (negative punishment). Are you still with me? A good way to think about it is that you use punishment when you want to decrease an undesired behavior, while you use reinforcement when you want the behavior to happen again.

Just like with reinforcement, there are two types of punishment: positive and negative.

Positive punishment: Positive punishment works by *presenting* an aversive consequence after an undesired behavior is exhibited, making the behavior less likely to happen in the future. For example, my chicken remembers that when she walked across the driveway, it was extremely hot. The next time she approaches the driveway, she walks alongside it, on the grass. Another example is that a dog, wearing a shock collar, barked at the neighbor and got shocked.

Negative punishment: Negative punishment happens when a certain reinforcing stimulus is removed after a particular undesired behavior is exhibited, resulting in the behavior happening less often in the future. For example, my youngest son had an attitude before dinner, so he didn't get ice cream after dinner.

Research with humans, dogs, and just about every other species shows that positive consequences are more powerful than negative consequences.

If you love your chickens and enjoy spending time with them, training will be fun!

Why Train a Chicken?

So, why in the world would you want to train a chicken? Simply put: it's fun and can put a smile on your face!

My guess is that, if you have already purchased this book, your level of curiosity is as high as mine. It is very interesting to me how these little creatures learn and how much we can do with them—it almost seems endless! I would say that you can probably teach chickens to do almost as many tricks as dogs can do. Chickens are curious, cunning, mischievous, and extremely smart, and, when you gain their trust and interest, they will stick with you. Almost every time I put one of my chickens on a table for training, she willingly stays on the table and engages with me. I don't clip my chickens' wings, so they are free to fly away at any moment (and don't get me wrong, sometimes they do!), but most of the time, they stick around, waiting to see what I'll do next and how they can use their brains to get the treats that are waiting for them.

I love watching someone learn how to train chickens. One of the neat things about training these birds is that there are no preconceived notions about how to train them. We all have ideas about how to train dogs, horses, and even some other animals. But, when you are getting ready to train a chicken, you are mentally starting from scratch.

When someone asks me why a person would want to train a chicken, I sometimes feel like a fool because I could just go on and on about it! So, now that I have your attention, let me tell you. I first thought it would be fun to train a chicken because I have been a dog trainer for more than twenty-five years and I absolutely love to observe behavior. I observe behavior in people, dogs, and just about any other creature out there. I have a thirst for knowledge. In addition to dogs, I've trained cats, horses, and pigs—and chickens were next on my list! (I've also tried training husbands, which doesn't work as well, but positive reinforcement goes a long way!)

So, who would enjoy training a chicken?

Dog trainers highly benefit from training chickens because it helps them fine-tune their observation skills and timing. Chickens are super fast, so working with a chicken will teach you to notice even the smallest of behaviors. And you end up realizing how slow humans are by comparison!

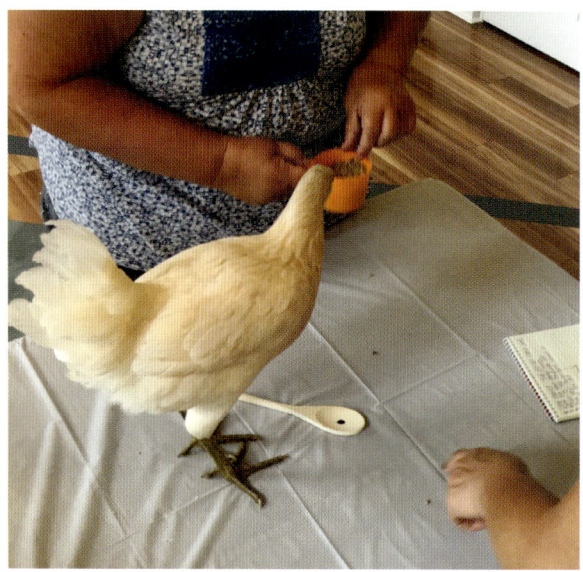

The interactive workshop allowed participants to get hands-on with chicken training.

Backyard chicken fanciers have a great time training chickens. I love animals, so when I can find a way to engage with my animals, it makes my relationships with them even more rewarding. Sometimes, we really want to do things with our animals, but we just don't know how to start or what in the world to do.

Corporations can benefit from training chickens. What? Yes, you read that right! It might sound a bit far-fetched, but hear me out. I recently taught a hands-on chicken-training workshop for a national insurance company. It was a team-building workshop for a department that has many out-of-state staffers who don't

Me and my expert chicken handlers (left to right): me, Jessica Cady-Bartholomew, Stephanie Weis, and Liz Perry.

get to see each other very often, so they brought in these team members from around the United States to attend the workshop. At first, people were greeting each other but not really talking to each other, but the atmosphere changed dramatically during the workshop. By the end of the seminar, they were enjoying themselves, laughing, talking comfortably, and having a great time.

Nursing-home residents enjoy interacting with chickens and helping them practice their skills, and you can find plenty of videos on YouTube or Facebook to prove it. Chickens are making their way into nursing and assisted-living homes as fantastic therapy animals. They seem to adapt quite quickly. Once they are used to the situation, they are fairly calm and quiet. The residents really enjoy the chickens' visits, and it's rewarding to put smiles on everyone's faces by showing off what the chickens can do.

Schoolchildren love visits from chickens and the opportunity to work with the birds. My son and I have taken our chickens into schools many times. Even though we live in a rural area, there are still many kiddos who have not seen chickens close-up. The children are curious, and the chickens seem to enjoy showing off their skills.

My son Corey and I at a school visit to teach the kids about chickens. Children and chickens alike had a great time!

My son Rudy and his chicken, Sam, during a chicken training workshop.

Chapter 2

Understanding
Your Chicken

We have a number of chickens (which is always about ten more than my husband thinks we have!), and it's easy to understand their different personalities once you get to know them. Chickens, like other species, including humans, have different personalities when they are with their flocks versus when they are away from their flocks or in new situations. I'm sure I would act differently with my best friend than I would if meeting the First Lady of the United States. That is part of the reason why I like to bring a chicken out of the flock to work with her a bit and get to know her before I start training.

While each chicken is an individual, you can safely categorize some chicken breeds into different "behavior" groups. Of course, everyone has his or her own experience, but here are some of my observations:

- **Sweet and calm (generally heavier-boned chickens):** Australorp, Orpington, Rhode Island, Sussex, Welsummer
- **Kid-friendly:** Silkie
- **Flighty (more nervous):** Andalusian, White Leghorn, many bantam breeds

If you're looking at different breeds strictly for training, you might want to stay away from the crested breeds. My guess is that a crested chicken might not be able to see the treats that you're offering her with all of those feathers in her eyes. Or, if you choose a crested bird for training, you might want to trim her bangs just a bit.

The Andalusian is my favorite breed to train, probably because this was the first breed that I ever really connected with. A splash Andalusian, Feathers (appropriately named by my youngest son), would trot

Corey's favorite chicken is CeeCee, a kid-friendly Silkie.

over to you and sit right on your lap. She enjoyed being with people, and we enjoyed being with her. She amazed me with her intelligence. She truly was a "thinking chicken." Any time I offered her an opportunity to learn, I could just about see the wheels turning in her mind. She wouldn't walk away, and she wouldn't jump into something without thinking it through. She would look at what I presented to her, look at me, and then make a calculated decision. I could actually see her thinking about what she should do!

Feathers and I at a demonstration at Cluck the Chicken Store.

Sam was a blue Andalusian. So far, she is probably the smartest chicken I've worked with. I could teach her something in one session that would have taken a smart dog at least three sessions to get—and she was rock solid, too! After taking a year off from training, and just being a chicken in the backyard, I took her to my son's school. I had all of my supplies ready, but I had no clue as to whether she was going to remember any of her training. When I put her on the table—*bam*! It was like we had never taken a break from training. I was impressed.

The more time you spend with your chickens, the more you will be able to understand them and help set them up for successful training sessions. For example, I know my chickens like to lay their eggs in the late morning. So, if I take them somewhere, I am going to make sure that we are back home so that they can go into their nesting boxes when they like. Or, if you have a chicken that thinks your family dog is a coyote who's out to get her, be sure to put the dog away when you're training.

Signs of Stress

Stress is something that can take over the mind as well as the body. Each chicken will have a different level for tolerating stress. Just like with people, some may become stressed out easily, and some may be more like ducks and let things roll off their backs (pun intended—for all of my duck-loving friends). You'll need to observe each chicken to help create an environment that is less stressful for her. Signs of stress include becoming withdrawn and distracted, disinterest in the treats and what you're trying to do with her, looking nervously around, wiping her beak, or trying to fly off the table. Longer-term stress includes loss of appetite and overpreening/overplucking her feathers.

Aptly displaying the chicken's amazing memory, Sam retained all of her training even after taking a year off.

With that in mind, I like to put my chickens in training situations that are *mildly* stressful to help them learn to work through potentially stressful situations. I've found that chickens are actually pretty darn adaptable, so creating slightly stressful situations will help your chickens become more tolerant and less stressed.

For example, say you're going to take your chicken to the fair. You shouldn't wait until the day of the fair to have her experience a variety of things for the very first time. For the chicken, going to the fair can include having a bath, being enclosed in a crate, going on a car ride, arriving someplace other than the backyard, and having to be crated next to a chicken she has never seen before. She could be in a building that smells different, sounds different, has birds other than chickens... And, by the way, she is by herself, away from the safety of her flock. That might be a bit stressful for her, eh? Most chickens are pretty adaptable and will become perfectly comfortable after a couple of days at the fair, but, because we love our chickens, we want to help make life easier and less stressful for them.

Many of my chickens go with me to events, seminars, workshops, and even friends' houses. I like to introduce them to new situations slowly. If you know that your chicken is going to be

During a team-building seminar, Pearl had other things on her mind, like finding an escape route.

headed on an outing in a week, start working with her now to help her be more relaxed. Take a few days to get her used to the crate—or, if you haven't really handled your chicken much, you'll want to start there. Spend a couple of days holding her and developing a level of trust and then have her go into the crate. Try feeding her in the crate. Teaching her to come when called is very useful and will help you catch her in the barn without added stress.

A couple of summers ago, when I took my hens to a workshop in Madison, Wisconsin, I had carefully chosen which five hens I was going to bring. They were all the same age, but they had very different personalities and

Acclimate your chicken to a travel crate before you have to use it.

learning ranges. Some learned very quickly, while others needed extra time to trust people and their environment.

The girls had all been crate trained and clicker trained, and they had been in the car with me just once prior, to help them get used to the movement of the vehicle. Each chicken had her own individual crate, so even though they couldn't see each other in the car, they could hear and smell each other and fully knew that they were not alone. (Sometimes I think it's easier to transport just one chicken at a time, though. If four chickens are perfectly fine in the car, but one chicken is panicking, that one chicken will cause the others to panic as well.)

On the morning of the seminar, I woke up early because I had an hour-long drive, and I wanted to make sure I had everything ready. I had recruited some of my best "chicken-lady" friends to help me once we got there; this was a really big day! I was feeling a little nervous, which I'm sure didn't help. But the worst part was the weather that I had woken up to. The sky was as black as night, and the trees were just about sideways because of the wind. I knew I couldn't cancel because there were so many people involved, so I packed my girls in the van

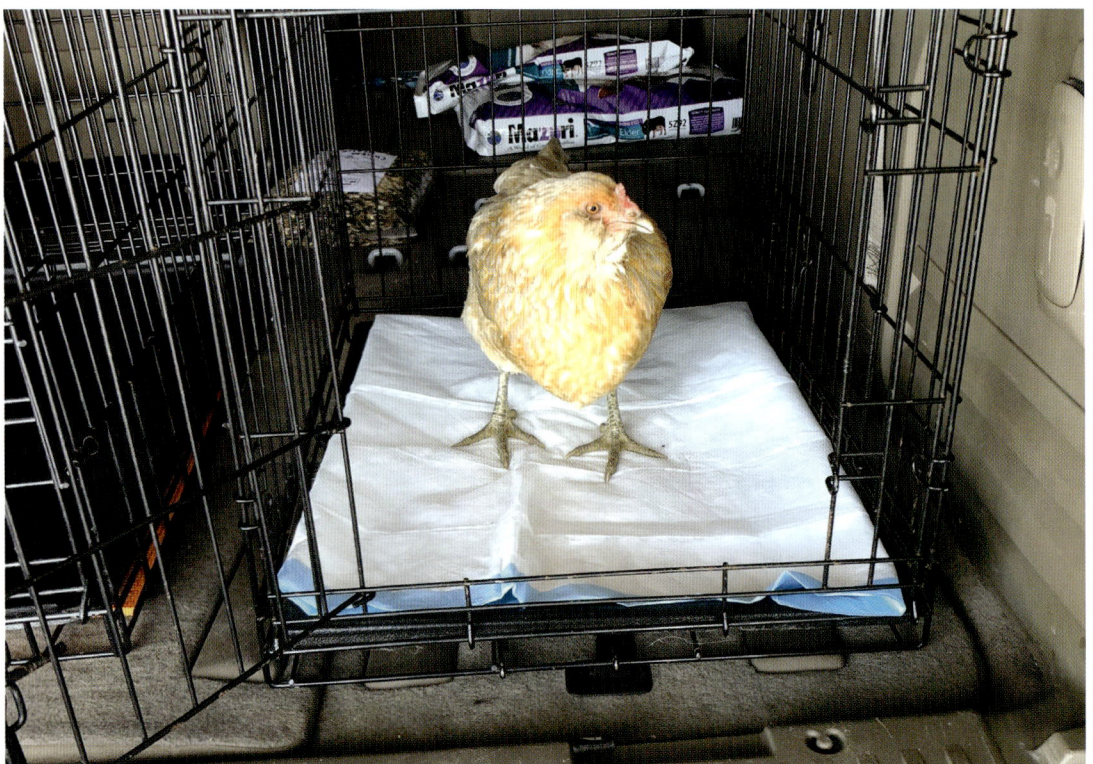

A crate will give your chicken a sense of security during car travel.

with me, and away we went.

I was hoping I wasn't going to be late. The drive took an hour on a regular day, but now, with the storm, I was getting worried. I wanted to get there before everyone else so I could have extra time to set up and get the chickens acclimated. On the way, I got a couple of phone calls from my friends, and they were running late, too. My blood pressure was rising, and I was reaching my threshold.

The wind was rocking my vehicle from side to side. My friends and I had to run through the rain to transport the chickens in their crates (one by one) from the van to the building. Thank goodness we all arrived in one piece. I was probably a shaking mess at that point because, first of all, I hate storms. Plus, I was so nervous about how the chickens would react after all of the commotion. They had only been in the car once and had only been to a place other than home once. Now I had brought them through a storm and into a brand-new building with air conditioning and a lot of voices they'd never heard before. On top of all this, I was expecting them to "perform" for the workshop. Talk about setting myself—and the birds—up for failure!

Well, to my thankful surprise, the birds acted exactly like I thought they would. Two of them were rock solid (just as they had been in their training at home). Two did fairly well, considering the circumstances. And one was completely flighty and wanted nothing to do with the team that was working with her, the treats, or me. She just wanted to hop on top of her crate and survey the ceiling to see if she was able to jump or fly high enough to perch somewhere—just as she had during our training sessions.

Threshold

I mention the word *threshold* in relation to training. Stress is a little different. Threshold is when there are too many things going on in your bird's environment for her to really concentrate on training. I had definitely hit my threshold with that seminar, but after seeing my friends and knowing that my birds were all OK, it quickly wore off.

If chickens hit their threshold, they are unlikely to take treats. I like to start my first training session with a chicken by just spending time with her and seeing what she enjoys eating.

Little Red wasn't in the mood for puppy Axl's curiosity.

Sometimes the chicken won't take a treat on the first day because she is too nervous. Chickens spend 99 percent of their time in their coop, with their friends, so when you take them out of their coop, the world can be a bit scary.

I admit that I hit my threshold often. I have three kids, three dogs, and a very busy lifestyle. Here's a look at my threshold: I wake up late, which means that my youngest son will need extra management to get ready for school. I get him up and moving, only to discover that we forgot to do his reading and sign his notebook the night before. Ugh. Then my older son gives me a permission slip that should have been signed two days ago. Two minutes before we have to leave, my daughter reminds me that she needs ten dollars for a fundraiser, and of course I don't have any cash in my wallet. Sigh… So I have to write a check, which is the last thing I want to do when we are running around before school. Snow pants, lunchboxes, backpacks, check! We are finally out the door. After I drop off my kids at school, I realize that, in the craziness of the morning, I forgot my own notebook that I need for work. Then my husband calls and asks me what I'm making for dinner—is he kidding? I blow up at him over the phone. I wasn't really upset with my husband, and if he would have asked what was on the menu for dinner two hours later, I probably would have been just fine. But I had hit my threshold. I had too many things going on at once that morning. We sometimes hit our thresholds, and our chickens do, too.

If you notice that your chicken isn't progressing as quickly as you'd like, think about what is going on around her and if she may be at her threshold. Sometimes it may be something simple, like a curtain blowing in the wind or a family dog lying at your feet. Chickens are very aware of their surroundings. One thing I've found, though, is that once you have worked with a chicken a few times, she really does tend to focus on you and what you are working on with her.

Be patient with your chickens. They may not initially understand what you're asking of them, and it might be easy to get frustrated with them. Take your time! They are not in a rush to learn, and your main goal of training is really just to enjoy being with them. They'll get it—just take baby steps.

Socialization

Chickens need socialization just like any other creature would. I'm not a fan of immersion socialization (doing too much socialization at one time). Instead, I like to take things slow and make sure that my animals are left with a positive association each time.

You'll want to introduce your chicken to new sounds, smells, sights, and people. With puppies, Dr. Ian Dunbar—renowned veterinarian, animal behaviorist, and dog trainer—suggests that they meet a hundred different people by the time they are eight weeks old and then another hundred people during their first month with their new families. Dr. Dunbar also points out that

the socialization process shouldn't end in puppyhood.

What about chickens, then? If you want a really well-rounded chicken, I'd say go ahead and follow the same guidelines. As long as you are providing your chicken with structured, safe, and positive interactions, it's certainly not going to hurt! Expose your chicken to different types of people: tall people, short people, people with loud voices, people with beards, people of all ages—all sorts of people. Here are some ideas for socializing your chickens.

- Have one or two friends over at a time to meet your bird.
- Allow your bird to see a (very well-trained) dog. Have the dog crated so the bird can freely walk up to the dog but then retreat if she feels frightened.
- If you've trained your bird only in the family room, take her into the bathroom, the hallway, or another area in your house.
- If your bird has been only in the backyard, take her (safely) into the front yard.

Once your chicken is ready, take her on outings. Go to a pet-friendly location other than your house, such as a friend's house or a pet store. Be aware that your chicken will be a novelty, and people will probably want to meet her. Remember, she's relying on you to keep her safe, so take this into account when letting people greet her. If you see someone walking with his or her dog in the pet store, you may just want to walk down a different aisle.

Our chickens are exposed to all kinds of farm animals.

I love to take my chickens to my child's school and my grandmother's nursing home. Kids and the elderly are great socializers for animals. They are (usually!) well-mannered, curious, and happy to get a visit from a chicken. You can easily set up structured greetings so you can greet each student or resident one at a time. Make sure that you're aware of your chicken's emotions. If you notice that she's getting stressed, either remove her (put her back in her crate) or just give her short mental and physical breaks.

Whenever you take your chicken outside of your own safe environment, be sure to confine her. A chicken can easily be frightened and run—and we all know how difficult it is to catch a chicken on the run. Keep her safe and secure in a crate (plastic cat carriers or small dog crates work great). You can use a chicken harness if you are sure that there is no way she can escape from it. Just holding the chicken in your arms may not be enough, even for the most well-trained and bonded chicken. Remember, it's your responsibility to preserve your chicken's trust in you by keeping her safe. When correctly socialized, your chicken should become more confident, bonded to you, and less likely to shut down or try to get away from you and the environment or situation she is in.

Corey and I do a school demonstration with our Barred Rock, Checkers.

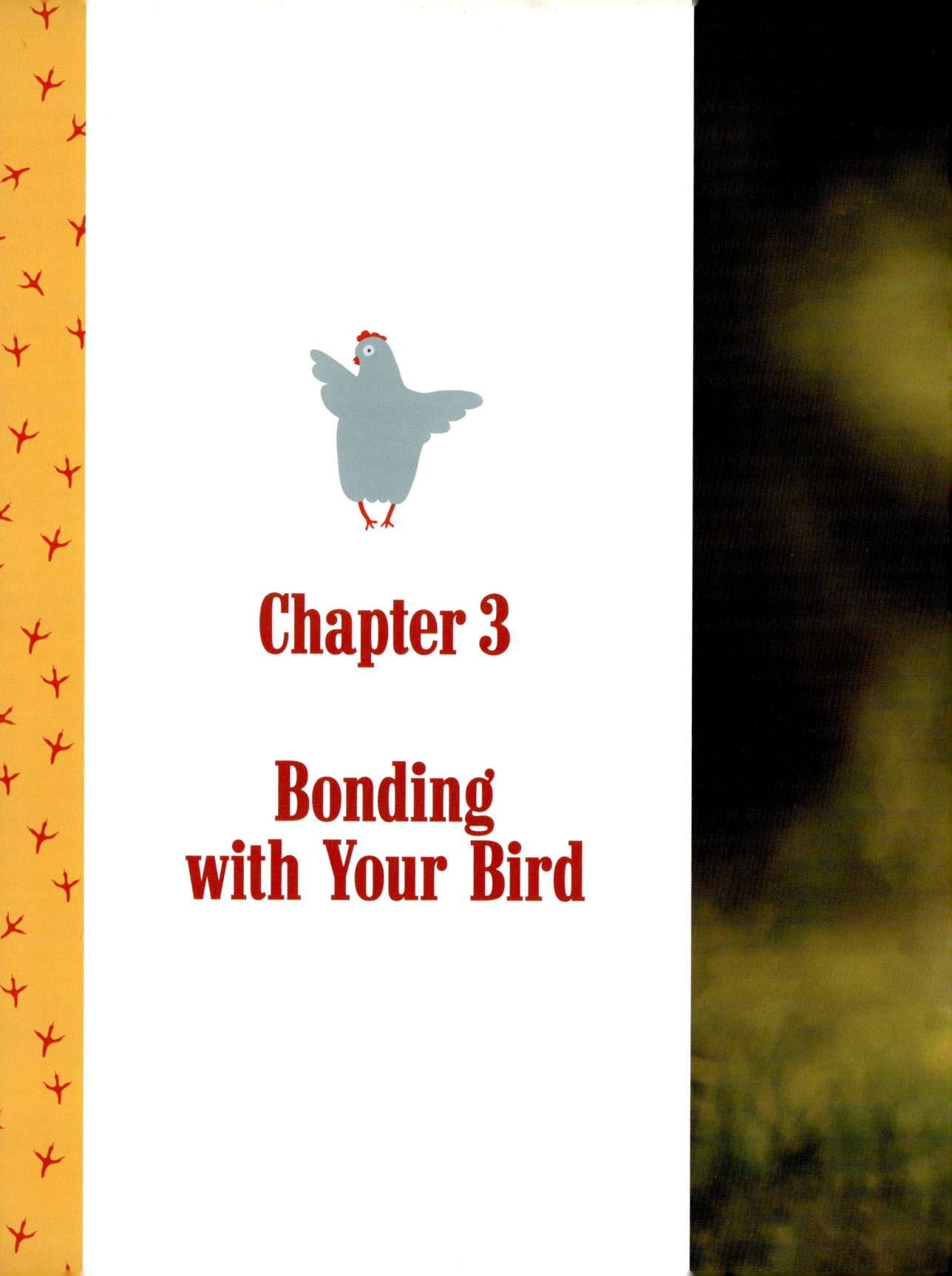

Chapter 3

Bonding with Your Bird

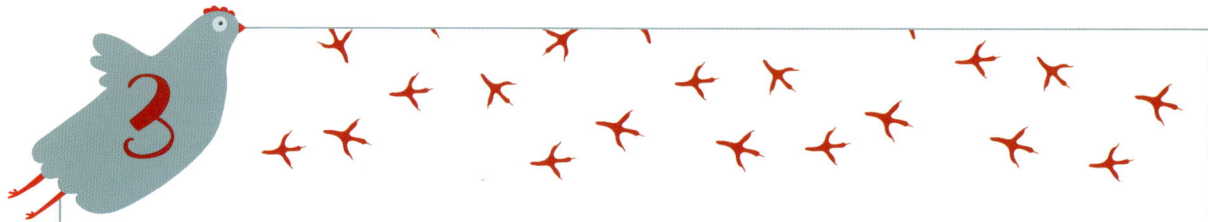

I t may sound silly to have to explain how to bond. But when it comes right down to it, bonding and gaining trust are very important if you want your chicken to work with you. Some people, like my middle son, have a natural way about them, and animals just seem to want to be near them. Others have to consciously work toward having relationships with their animals.

Early Socialization

I like to start handling my chicks when they are just a few days old. Day-old chicks are pretty fragile, and, if you're like most people, you're probably getting your chick when it's between three and five days old (from a feed store, shipped from a hatchery, from a swap meet, or from a friend). I've gotten my chicks from all of these sources as well as hatched my own. If you don't care if you're getting pullets (young hens) or roosters, hatching your own chicks is fun and rewarding—putting them in the incubator, counting down the days, spending time turning them each day, and candling them to see how they're coming along. Your chicks can get used to you

Our chickens are part of the family.

before they even come out of the eggs! Just like unborn human babies, chicks can hear your voice while still in the eggs. Even though we leave them alone for a day or so after they've hatched, the chicks still see you peering into the incubator and hear you talking to them. I will often start with my little chicken call— "chick-chick-chick"—while they are drying off in the incubator. At this point, they don't have anything to associate it with, but I hope that just hearing my voice will help with bonding.

I have five-day-old chicks that we hatched at home sitting next to me as I type this. I love listening to their little chips and coos. They are all pretty tired because they had a big outing today. We took them to my eight-year-old son's classroom for show-and-tell. They were a big hit! We were able to educate the kiddos about chicks, and the chicks got some great socialization. Just like with any animal, the more positive and appropriate handling and socialization they get at an early age, the better. Young chicks will have strong instincts, but they will also remember positive situations. For example, if you quickly wave your hand over their heads, they will most likely run away and chirp wildly, for fear that a hawk is going to swoop down. But if you bring your hand to a chick slowly, and that hand is usually offering a yummy treat, the chick will start to come toward you to greet your hand. The chick will have a positive association, rather than a negative, fearful reaction, to your hand.

I am excited to start working with these little ones, watching their personalities blossom, and seeing which ones will be my star students and which ones will need a little extra time. I have high hopes for them. Their sire is half Olive Egger and half Andalusian, and their mom is a Barred Rock. Barred Rocks are quickly becoming one of my favorite breeds. They are highly intelligent and have a good mental sense of balance about them. They don't jump into decisions; instead, they think about it for a moment, which I think helps them retain what they've learned and use it in their decision-making process. Yes, chickens think about their situations and make decisions based on past experiences. Isn't that crazy?

My son Corey and Sprinkles at a school visit, only about a week after we brought her home.

Once you've established a bond, you and your chicken can form a trusting friendship.

HANDLING YOUNG CHICKS AND ADULTS THAT ARE NEW TO YOUR HOME

If you picked up your chicks from a feed store, swap meet, or any other place where they have lived for a period of time, you'll want to allow them time to get used to your home. Chickens are highly adaptable, which is probably one of the reasons they have survived for millions of years. They don't take long to understand their new surroundings, the energy and routines of the home, and so forth. But when chickens—whether chicks, pullets, or adults—have just come to you, everything from the smells in the air to the textures under their feet is different, so let them settle in for a few days before jumping into training and bonding.

How Your Energy Affects Your Chickens

One of the things that many of my chicken friends and I love is to simply be around our chickens. If you've had a chance to pull up a chair and just hang out with your little flock, you can understand how relaxing it can be to listen to their little sounds, watch them enjoy dust baths, or nap in the sunshine. I think it's so interesting to observe chicken behavior. I've been a student of creature behavior since I can remember. One of the things that we don't often realize, though, is that our animals are also studying our behavior.

You know how sometimes a dog or cat will simply adore a person? Or how a parrot might despise a new guest in the home? There are many reasons behind bonding and acceptance of new creatures to your flock, family, or pack. Every creature has likes and dislikes, and, believe me, our animals can read us like a book! They are masters at observing our energy and making those "first impressions" (often better than we can). This is just my own experience, but I think that animals mostly feel comfortable with calm humans. When you introduce chickens to each other, you don't hold them face to face and make lots of excited noises, right? You may cage one of them and simply put her in the same area as the other chickens, allowing them to take time to get to know their new friend. Chickens are extremely curious creatures, so they will look at each other. They'll go up to the cage and peer in, maybe even peck around, but then they will most likely go on their way, ignoring the caged chicken (unless they're jerks, like a couple of the chickens in my barn, who purposely try to bully and torment the new chicken). We are kind, chicken-loving people, so we are going to take our time with our chickens and help make them feel comfortable around each other and around us.

Here are some ways you can use your energy in a positive way so that your chickens feel more comfortable around you:

Be calm around your chickens to help them feel comfortable with you.

- Don't walk in slow motion as if you're going to pounce on your chicken at any given moment.
- Do walk slowly but purposefully.
- Don't act like a child and erratically go from a standstill to a full-out sprint.
- Do move with calm, fluid motions.
- Don't flail your arms about. This is hard for me because I tend to talk with my hands, and sometimes my whole body will move, depending on my mood. Waving your arms around will make your chickens feel uneasy, though.
- Do try to keep your arms at your sides for the most part.
- Don't screech, yell, or laugh loudly at your own jokes. This is a tough one for me, obviously, but trust me—your chicken doesn't find it hilarious.
- Do speak in an "inside voice." Try to keep your speech less animated.
- Don't go a mile a minute.
- Do take your time. A chicken has all day to do basically nothing. Slow it down a bit. You may have your agenda and your goal for your training session, but try to relax. Enjoy the moment and soak in how calming it can be.

How to Start Bonding

So now you have your chickens—hooray! It doesn't matter if they are three days old or three years old. You are going to start bonding with them the same way. Depending on where your older chickens came from, it might take them a little longer to bond with you, but don't get discouraged! As long as you take your time, I'm confident that even an older hen that never really had human contact before can still be a fun bird to train!

Bonding: Level 1

If your bird is outside, grab a chair or a blanket to sit on. Take some yummy chicken treats along with you and just sit. Since chickens are curious by nature, it will only be a matter of time before your chicken decides to come near you. You can sweeten the pot by gently tossing a few treats close to your feet. If your bird is very skittish, toss the treats a little farther away from you. You want to figure out her comfort zone. If your bird wants to stay about 10 feet away from you, start

tossing your treats that far away (you may find that you have to work on your treat-tossing skills!). Once your chicken gets to that 10-foot range, start tossing them 9 feet away. You want to work on getting closer and closer. It may take you a few sessions, but you should work up to at least having your chicken come up to your feet to take the treats that you've tossed.

Remember your energy during this time. Don't be talking on the phone, and heaven forbid you sneeze loudly when your chicken finally feels comfortable enough to come close. (You may have to start over from the beginning if that happens!)

Bonding: Level 2

Level 2 should start where level 1 left off, with your chicken getting close enough so that you could reach your hand out to touch her. (But don't actually try to reach out and touch her right now; she will probably scurry out of your reach.) If you're sitting on the floor, continue to drop treats near your feet and then on top of your shoes and around your legs. (Only go into your chicken area barefoot if you want your toenails to be pecked, especially if you're wearing shiny nail polish.) If you're sitting on a chair, put some treats in your hand and slowly extend your hand toward your chicken. Sometimes placing your hand on the floor— rather than having it hover a foot above the ground— helps. Remember, chickens are used to pecking at the ground.

When your chicken gets the courage to peck at the treat from your hand, you're going to want to jump up and have a party, but don't do it! Right now, you're like a zen yoga master—calm. And don't be surprised if it hurts a little. Well, I shouldn't say that it hurts, but some chickens will peck pretty hard. Some take treats gently, and some don't. Some will take treats and then wipe their beaks in disgust. They all have their own little ways of doing things. You'll hit some milestones in your chicken bonding and training, and this is one of them.

Bonding: Level 3

Sometimes, when you start a bonding session, you may have to go back to level 1 even if you've already progressed beyond it. However, it will take only a fraction of the time to get from level 1 to level 3. Chickens have fantastic memories, so even if you go weeks or months between sessions, you may be pleasantly surprised at what your chicken remembers.

Once you reach level 3, you shouldn't have to toss your treats away from you to get your chicken to come near you. You may have to drop a couple of them on the floor, but at this point your chicken

TRAINING NOTE

Remember, these levels are general guidelines. Each chicken, each flock, and each trainer's energy is going to be different. Some of you may skip right to level 3 on the first day, while some may take a couple of weeks to even get through level 1. Be patient. If you have just brought an older chicken home, you don't know what she's been through. People may be pretty scary to her. Give her time to gain your trust.

At this point in your bonding, your chickens should willingly approach you.

should willingly walk over to you when you enter the coop area. Unless I'm training my chickens and they are in my house, they live in my barn, so I tend to refer to them in this book as "in the barn" rather than "in their coop area."

When I walk into my barn, I am greeted by my flock. If you are new to chickens, you might actually be scared by this at first—they literally bum-rush me at the door! I've conditioned my chickens to associate me with food, water, and treats (super-awesome treats!). They love me more in the winter, when there are no bugs for them to find, but that association carries over into summer as well, when the bugs are plentiful. It's a pretty cool thing to walk into your barn and have all of the chickens drop what they're doing and run over to you as fast as they can.

At this point, you should be able to walk into the barn and find that your chickens are happy to greet you. Drop a couple of treats near your feet and then hold your hand out, offering the treats now only in your hand. You might have to stand there like a statue for a while because your chickens may decide to "wait it out" and see if you'll just toss the treats instead. Once they figure out that you are not tossing them anymore, they should become brave enough to just walk right up to you and take them out of your hand.

Operant conditioning tells us that once an animal is doing a behavior consistently (at least 80 percent of the time), we no longer reward that behavior, and we start rewarding the behavior that we want. So, if I started with wanting my chickens to come within 10 feet of me, but my desired end result is that they walk all the way up to me and take treats from my hand, I am going to stop tossing treats for them when they are 10 feet away. Baby steps.

GUIDELINE FOR BONDING SESSIONS

- Sit in the barn and toss treats about **10 feet** from you.

- Once the chickens are comfortable taking the treats **10 feet** away from you, stop doing that and toss the treats **8 feet** away from you.

- Once the chickens are comfortable taking the treats **8 feet** away from you, stop doing that and toss the treats **6 feet** away from you.

- Once the chickens are comfortable taking the treats **6 feet** away from you, stop doing that and toss the treats **4 feet** away from you.

- Once the chickens are comfortable taking the treats **4 feet** away from you, stop doing that and toss the treats **2 feet** away from you.

- Once the chickens are comfortable taking the treats **2 feet** away from you, stop doing that and toss the treats **1 foot** away from you.

- Once the chickens are comfortable taking the treats **1 foot** away from you, stop doing that and **place the treats on the ground** next to you.

- Once the chickens are comfortable taking the treats from the ground next to you, stop doing that and **place the treats in your hand** and then **place your hand on the ground** next to you.

- Once the chickens are comfortable taking the treats from your hand on the ground next to you, stop doing that and **hold your hand (with treats in it) just a couple inches off the ground**.

- Once the chickens are comfortable taking treats from your hand a couple inches off of the ground, stop doing that and **raise your hand (with treats in it) to their eye level**.

- Once your chickens are comfortable taking treats from your hand at their eye level, **change your posture**. If you were sitting on the floor, kneel. If you were kneeling, squat. If you were squatting, stand up and bend over. If you were bending over, stand up straighter. Essentially, you want to change your body shape. Sometimes, with animals, if we change our body shape, we can become more intimidating, and sometimes they may even view us as a different person altogether.

Chapter 4

The Clicker

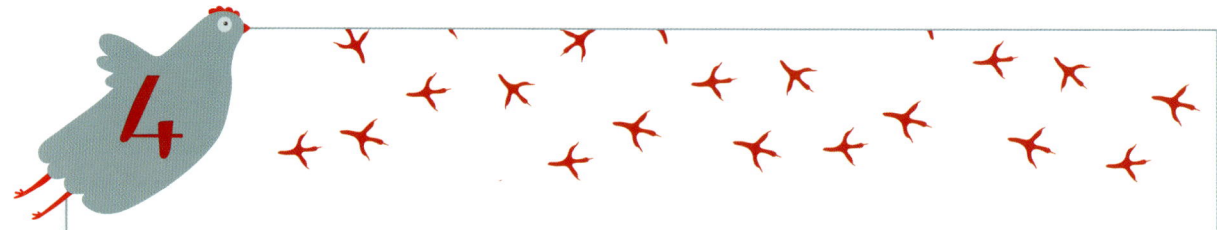

The Background of the Clicker

I'm guessing you've heard of Dr. Pavlov and the famous "Pavlov's dogs." Dr. Ivan Pavlov was a Russian physiologist and chemist born in 1849. While studying dogs and their digestive process in the late 1800s, he discovered a *conditioned reflex*. His study initially intended to research the link between salivation and the actions inside a dog's stomach. He discovered that the two actions were related by reflexes present in the dog's autonomic nervous system. If the dog didn't salivate, the stomach would not get the message to begin digestion. Pretty interesting, eh? Dr. Pavlov wanted to do more research to see how an outside stimulus would affect the process. He decided to ring a bell at the same time that he gave food to the dogs (and the rest is history!).

The clicker should be small and easy to use.

Up until that point, the dogs would begin to salivate when they actually saw the food and ate the food. Once Pavlov began to ring the bell right before feeding the dogs, the dogs started to salivate at the mere sound of the bell, even if there was no food within sight or smell. Pavlov referred to this as a *conditioned reflex*, which is basically an action that is learned. He called his bell-ringing process *conditioning*. He also learned that if he rang the bell too many times with no food present, the conditioned reflex would eventually become repressed. In other words, if he did it too many times, the dogs would no longer salivate when they heard the bell because they learned that it no longer meant that food was coming, thus it no longer had any meaning to them.

Dr. Pavlov's bell is essentially the same as using the clicker. The animal hears the click (bell) and understands that the reward (food) is about to come. I particularly love to use clicker training with chickens because they are so responsive to it. Horses, cows, and pigs all really enjoy the company of people; they like petting, hanging around with us, and playing with us. Chickens? Yeah, I know plenty of chickens that are the same way. But, when you're training chickens, they don't really feel like taking the time for petting. They don't really want to hear "whadda good chicken!" They want the treat, and they want it quickly!

Clicker training falls under the operant conditioning umbrella, and it's a great way to train your chicken. I like it because it does not involve force, and it allows your bird to think on her own. With this type of training, we shape and reward positive behavior with clear signals and reinforcement. The clicker is like a small children's toy (hand one over to my eight-year-old son, and you'll see what I mean!), but, when used effectively, it can have a pretty amazing effect on your chicken. Your chicken will learn to associate the small "click" with a prime reinforcer, like food. It's cool because it becomes a game for

Clicker training became popular with dog trainers, and the principles work well with many other animals, including chickens.

your chicken—she will try to see what behaviors she can offer to get you to click, and for her to get the treat! I especially like to use shaping (see definitions starting on page 72) and clicker training because it does not involve any type of compulsion or harsh manipulation to control the animal's behavior. If your chicken does something wrong, you simply don't reward it. Once the bird makes the connection between the click and the treat, you can begin to use the clicker as a way to reinforce behaviors that you like (operant conditioning).

There are many chickens that will come running when their names are called or may even know commands like "up" or "go in." But chickens, like our dogs and other animals, hear our voices quite often. Using the clicker is nice because the animal hears it only during training. It's also nice because it makes it easy for us to pinpoint a behavior. The moment your bird does something right, you click! Clicker training also keeps your chicken engaged. You can work with your chicken from a few feet away and help her focus on the behavior at that moment.

Have you heard of marker training? Marker training is basically the same as clicker training; the only difference is that instead of using the clicker as a reward signal, you use some other signal. It might be a vocal signal, such as "yes" or "good." When the basic concept behind this

The chicken will be engaged with training, waiting for the click to mark the desired behavior.

form of training was introduced to the general public by psychologist B. F. Skinner in a 1951 *Scientific American* article, he chose a clicking sound simply because it appeared to be more precise and was easily heard. However, according to Skinner himself, the marker doesn't even have to be a sound but could be some sort of visual signal, such as a light flash or a hand movement. I've seen people use laser pointers with chickens as both markers (in place of the click from the clicker) and targets. I like to use a laser pointer or pen flashlight in place of the click when I'm training deaf dogs; they can't hear the clicker, but they can associate the light as a marker when they have done something right.

Clicker training began to grow in popularity among dog trainers following the release of Karen Pryor's book *Don't Shoot the Dog* in 1984. Her workshops and videos emphasized the use of the clicker, and, after a while, dog trainers around the world—including me!—started using clicker training as their main method of training. I've found this to be the most effective way (by far!) for training our little *Gallus gallus domesticus*. For whatever reason, when we use our voice as a marker, we don't time it as well as we do with a clicker.

The Clicker-Training Process
Choose Your Treats

First, you need to find out what treats your chicken likes. Actually, you need to find out what treats your chicken can't possibly live without. I've used regular chicken feed before, and it's kind of "meh." If you have a chicken that really enjoys working, you might get away with using chicken feed, but I normally like to use high-value treats for training. There is a variety of treats that a chicken might consider "high value." Each chicken is different, of course, but remember that they are like little dinosaurs, so they will eat just about anything!

I'll use different treats for different things. If I'm going to the barn and want to give my chickens something fun, I'll give them all sorts of things, such as:

- Strawberries
- Melon
- Apples
- Corn on the cob
- Spinach
- Canned pumpkin
- Bagged-up "scratch seed" from the feed store
- Ground beef. I know that my vegan and vegetarian friends are not going to like this, but my chickens love it. It's my go-to treat if I really need them to be motivated; for example, if we are going to appear on TV or be in a large group of people. I know my girls will go berzerk for ground beef!
- Ground-up graham crackers (I recently discovered that Sprinkles loves these!)
- Mealworms. I like to toss mealworms around the barn for the chickens to find in the winter, but I don't recommend them as training treats. When you put them in the treat container, you'll notice that they're very slippery. When your chicken goes to peck at them, you'll end up with mealworms all over the table and floor. Learn from my mistakes!
- Pretty much any of your leftovers

When I'm training, however, I use more specific treats. I like treats that my chickens love and that I can offer in very small pieces (so I can give them a lot of treats and they won't get full too quickly). As I mentioned, my mini dinosaurs will do anything for ground beef. They also eat bugs all day (bugs = meat and protein). Now, I don't give them ground beef every day, and, when I do, I don't give them a lot. A little goes a long way.

Getting Used to the Clicker

By now, your chicken is used to you holding her, petting her, and generally being around her, so you can introduce her to the clicker. I like to train my chickens on a waist-high table that I bought in a department store's camping section. It's height-adjustable and large enough for a chicken to walk around on but small enough so that I can basically reach my chicken wherever she is on the table (she can't walk away from me). I think it's perfect for training chickens. Once

I have my chicken on the table, I will start to click and treat for any behavior I see—literally *any* behavior, including:

- Turning her head
- Taking a step
- Stretching her wing
- Stretching her leg
- Lying down
- Standing still (I reserve this for when I think that my chicken has lost interest, in which case clicking/treating her will usually spark her curiosity. Remember, chickens love to eat!)

Click and treat your chicken for doing any type of random behavior when you're first starting to teach her about clicker training. In the beginning, if you wait for your chicken to do a specific desired behavior, it might not happen, and then your chicken will end up losing interest. So, in order to teach her that the click means a reward is coming, you should start clicking her for *any* type of movement. The chicken will start to understand that each time she does something, she's going to get a click and reward. Once she understands that, she will start offering up behaviors in order to get the treat.

Clicking and treating often is also essential when you're working with your chicken. Chickens are super-fast, and if you do not give immediate reinforcement for a behavior, a chicken will assume that the behavior is not what you want, and she will either move on to the next thing or completely lose interest in you. So, when your chicken is on the table and you're ready to work, be ready to click and treat everything extremely quickly.

When you first start clicker training, click and treat for everything.

I advise keeping your training sessions to about a minute long. You really should keep your training sessions very short in the beginning because you don't want your chicken to lose interest, and you want to be able to give her frequent mental breaks.

Your Training Area and Equipment

Everybody has a designated chicken-training room right? Well, for me, I usually have my chickens in their crates right on top of the dog crates in my front room. Did I mention that I have a very tolerant husband and family? Because I don't have a designated area for chicken training in the house, I usually set everything up when I'm going to train and then put everything away when I'm done training. Fortunately, it doesn't take much equipment or money to train your chickens. You'll need the following supplies:

Table: I love my adjustable camping table. It's easy to transport, and it's easier for me to do training with my chickens when I don't have to bend down all the time. I also think that, because the table is a bit higher off the ground, the chickens are less likely to want to hop off of it.

Measuring cups: A measuring cup is perfect for holding treats. It has a little handle, and your bird won't make too much of a mess if you fill the cup only partway. If you fill it up all the way, you'll end up getting crumbs all over your table—and then your hen will stop her training to peck at all the goodies that just fell in front of her!

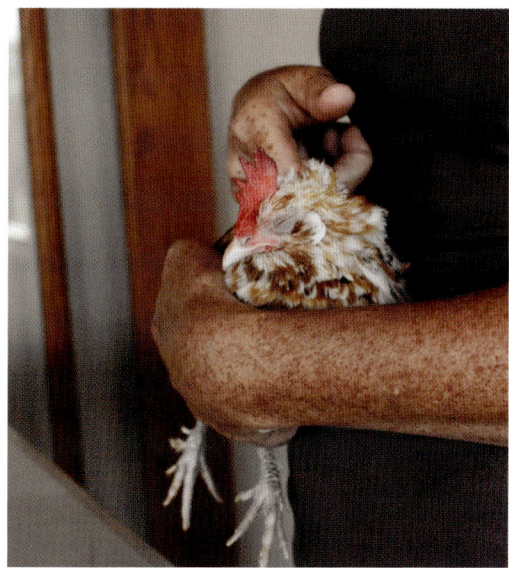

Training breaks will relax and reenergize your chicken.

A measuring cup makes a perfect treat cup, and you can attach the clicker to the handle to make click-treating easy for you.

Besides making the cup easy to hold, another nice thing about the handle on the measuring cup is that you can Velcro your clicker onto it. It's the perfect place for your thumb to land when you need to click. And, this way, you'll only be using one hand (clicker and treats in the same hand) rather than having to use two hands. I also like this method because I hold the measuring cup with treats behind my back, so when I click, the clicker noise comes from behind my back and is not too loud. I found my measuring cups at the dollar store, and I'm sure you can find them at resale stores, too—remember, you don't have to spend a lot of money if you want to train chickens!

Clicker: Nowadays, you can find clickers at most pet-supply stores and certainly online. I like the original type, which kind of looks like a children's toy, and the type that has a dial on the back so you can make the click louder or quieter. Some chickens are rather sensitive to sound; in those cases, you'll want to make the click much quieter. Remember, you want your chicken to be excited when she hears the click (because, to her, it means a treat is coming). You don't want her to be freaked out by the click.

Essential training tools include a clicker, a treat container, paper plates, a marker or pen to make dots on the plates, and an apron to keep yourself clean.

Apron or old shirt: We all know how much these little buggers like to poop! And, they love to embarrass us, so it's probably best to wear an apron or old shirt that you don't mind getting ruined.

Treats: As mentioned, use a variety of treats to find out what your chickens love. I like to use three or four different treats in my cup. That way, my chicken won't get bored and will be surprised each time she pecks because she won't know which treat she's going to get.

Paper plate: I like the small ones because they're easy to pick up and move around the table.

Be sure to have an apron to keep yourself clean during training sessions.

Pen or marker: I use this to make marks on the paper plates.

Paper towels or baby wipes: You can use paper towels or wipes and maybe vinegar in a spray bottle for easy cleanup. This works well, especially when I take my chickens on outings.

Hand sanitizer: If you're going to let other people handle your chickens, bring hand sanitizer for them.

How to Stand and Treat

It may sound strange, but there is a right and wrong way to stand when training your chicken. If you really want to do shaping with your chicken, you have to be careful that you don't inadvertently give her cues. It's very easy for us to accidentally give off body language that tells our chickens what we want from them. For example, if you have your paper plate with the dot on the table with your chicken, and your chicken is disinterested, you may inadvertently turn your shoulder or your hips toward the paper plate. Or you may look down at the paper plate. The chicken could perceive all of these little subtle body movements as cues from you. If you're doing shaping, you really want your chicken to figure out what behavior you're looking for on her own.

You may feel awkward in your movements at first, but you'll learn to develop quick yet fluid movement. You want to be able to treat your bird quickly without letting treats spill on the table. If you do occasionally get a crumb or two on the table, quickly wipe it off so your bird isn't distracted. Believe me, if she thinks that she'll get a buffet on the table each time she pecks at her treat container, she'll do it every time! And the last thing you want is a bird that makes a huge mess on the table simply so she can have an extra snack. Plus, she'll start looking at the table for rewards instead of looking to you for her rewards.

You'll want your click-and-reward to basically become an automatic response from you. The split second you see your chicken do what you want (which may not be the actual behavior, but a baby step toward the behavior), you must quickly click and reward.

I've found it helpful to record myself training, and you may find the same. Sometimes, the videos turn out great, and you may want to share them with your chicken-training friends. But the real reason for recording and watching your training sessions is to see how *you* are doing as a trainer. I can't tell you how many times during training I see my bird do something "click-worthy," but, by the time I see it and process it, she's already done something different. You have to capture the split second your bird does the behavior so you can click and treat.

Realizing that you are a slow human will be something that happens often, believe me. Prepare yourself before you watch a video of your training session because you'll probably be able to see a whole heap of mistakes that you made. You'll notice human errors like timing incorrectly, standing incorrectly, baiting your bird by moving your shoulder down, not realizing

Make the click-treat an automatic, fluid movement.

that she is about to fly off the table before she does it, and so on. Yes, it's a bit humbling, but it's really worth it if you want to become a better trainer.

When I was training my black lab, Buster, I inadvertently taught him to sneeze (I know—what a goofball!). I didn't even really understand what was happening until about two months later,

LET'S CONNECT!

If you post training videos on social media, be sure to tag me; I'd love to see them! My social media information is in About the Author at the end of this book.

when I had him at a training facility and we were working on a heeling exercise. We would be walking, and then all of a sudden he would look up at me and sneeze. Then we would take a few more steps, and he would look up at me and sneeze again. I kept looking up at the ceiling, wondering if maybe there was dust in the air or if, when he looked up at me, the bright lights on the ceiling caused him to sneeze.

Being the slow human that I am, it took me a while to realize that at some point earlier in our training, Buster must have sneezed, and I must have looked down at him. *Bam*! It was basically like I had clicked and treated him for sneezing. By simply turning my head down to look at him for sneezing, I had immediately given him a reward. Attention is definitely a reward. If your dog is barking, and you tell him to stop barking, you've just given him attention for barking. Buster learned to sneeze in order to get attention. I think he was saying, "Hey lady, you haven't treated me in two seconds and I'm ready for another!" Oh, Buster...

So, understanding how important body language is when you're working with your chickens is extremely important. Chickens are very visual, and they're *way* faster than we are. You want your chickens to be thinking and problem-solving when you're teaching them behaviors. You want them to learn on their own without your guidance (or cheating). You don't want to have to guide them, lure them, or inadvertently show them what they should be doing. They're smart enough to learn on their own (unless you're working with Silkies, which are so sweet but may take a tad longer to learn behaviors). In general, chickens are super-smart, and they're going to figure it out on their own time; we just have to let them.

When you're doing your training, I want you to think about being a little robotic. You're going to stand extremely still, you're going to breathe with your mouth shut, you're not going to talk, and you're going to hold your treat container behind your back. The moment your chicken does something that is "click-worthy," quickly yet gently reach forward and offer her the treat right under her beak. Don't make your chicken move toward you or walk away from whatever she's doing to get her treat.

If you are about to give your chicken numerous treats within a few seconds of each other, hold your treat container close to your body rather than putting it behind your back each time. Remember, you must be able to move quickly! Hold both hands on the cup when it's close to your body (to hide it from your chicken) and then reach forward with one hand to reward.

Click when the cup is next to your body, move your arm with the cup quickly right under her beak, allow her to take two or three pecks, remove the cup and bring it next to your body. Remember to give your chickens frequent mental breaks. All of this thinking can be exhausting!

Work on being quick and having a high level of reinforcement when you're training. If you click and treat quickly and often, your bird will match your pace, or training style. If you are slow and miss cues or behaviors, your bird will not learn very quickly. For example, if you're working on teaching her to peck at a dot on a paper plate, and she keeps looking at the dot but you are not clicking for it, she won't understand that the dot has any correlation to what is expected of her. Or, if she looks at the dot and then quickly looks at the door, and then you click, she'll think that looking at the door is what gave her the reward. You may still get to the end result of her pecking at the dot on the paper plate, but it will take you much longer. Your bird will become more distracted, maybe even frustrated, and won't learn as quickly. So be sure to have a high level of reinforcement at a quick pace.

Rate of Reinforcement

Because the click is what tells the bird that a treat is coming, click and then treat immediately in the beginning. After your bird is a clicker-training pro, you don't have to provide an immediate reward after the click. Remember, the click is not the reward; it simply tells the bird that a reward is coming. After your chicken truly understands that concept, you can start giving the treat a moment or two after the click.

I introduce "delayed gratification" so my birds can work on impulse control. If, throughout my training, my bird always expects the treat *immediately* after she hears the click, she will become impatient if she hears the click and the treat is not right in front of her beak. So, once she knows what the click means, I will start to offer the treat a split second later than I had been (remember, they're still much faster than us!).

Exiting the crate calmly is a good delayed-gratification exercise.

An example is something as simple as coming out of the crate when the door opens. If a bird is in her crate, and the door opens, that is a reward (being able to come out of the crate). Some birds will "rush" the crate door and come barreling out, but I like to teach my girls to wait a second before exiting the crate. It's good for the chicken's general manners and safety (she can't run out and try to fly away), and it gives me a chance to get ready to pick her up. It also sets a tone of patience and calmness for training.

If the crate door opening is a "trigger" for my bird to rush out, but I want her to wait calmly in the crate for a moment, then I am going to reward her for the calm behavior. I want the action of the crate door opening to mean that she stays in the crate for a moment, so I'll take baby steps in rewarding the calm behavior.

1. Put your hand on the crate door and treat (you'll have to reach a pinch of a treat through the crate door with your other hand).
2. Repeat step 1 three to five times.
3. Put your hand on the crate door and unlatch it, but don't open it, and treat.
4. Repeat step 3 three to five times.

5. Put your hand on the crate door, unlatch it, open it about half an inch, and treat through the slats in the door (not through the opening you've created).

6. Repeat step 5 three to five times.

7. Open the crate door about an inch and treat through the slats in the door (not through the opening you've created.

8. Repeat step 7 three to five times.

9. Open the crate door about halfway, continuously treat through the opening for three to five pecks, and then quickly yet gently close the door.

10. Repeat step 9 three to five times.

11. Open the crate door all the way, continuously treat through the opening for three to five pecks, and then quickly yet gently close the door.

Your bird will learn that when the door opens, it is to her benefit to stay put for a while because she is rewarded for her calm behavior.

Another example to teach delayed gratification is with one of the first tricks that you will teach your chicken: pecking at a dot (see chapter 6). Once your chicken knows this behavior inside and out, start to wait a moment before treating. You will still click immediately as she is pecking at the dot but then pause just a moment before treating her. Remember—your idea of a moment and your chicken's idea of a moment are probably very different! Start with just half a second and work your way up to a couple of seconds. You want to be quick enough so your bird still knows that she's getting a reward for that behavior. If you wait too long, she may start going

Sprinkles is very comfortable in her indoor crate.

through her repertoire of behaviors, trying to figure out how she can get that treat, but this isn't an exercise in shaping behavior—it's an exercise to teach impulse control.

So, how do you know when your bird understands what the click means? Sometimes, when I'm first training a bird, she will try to get at the treat container, just like a puppy would try to get a treat out of your hand. But, as you progress in your training, your chicken will understand that the treat comes only after she hears the click. When you see that your bird is no longer trying to get treats just because she sees that you have them, but instead looks to the treat container *only* after she has heard the click, you will know that she is associating the click with the reward.

Like your timing, your rate of reinforcement in the beginning of training, or when you're training a new behavior or trick, should be rocket fast. After your bird understands what behavior is expected of her, or if you are working on a behavior in which you want your bird to slow down or stand still, start to lengthen the rate of reinforcement. For example, when I crate-train a chicken, I will gradually wait a little bit longer each time after the click to offer the treat. This enables my hen to understand that settling in the crate and not moving is what earns her the treat. I still want to click to let her know that she is doing something right, but there is delayed gratification in actually getting the treat. If she were to stand up and walk away, she's not going to get the treat. But as long as she is sitting in the crate calmly, she is going to get the treat, even if it's moments after she has heard the click.

Here are some situations you should look out for:

Jumping off the table: When you first start, it may be a good idea to have your table up against a wall. That way, the wall is on one side of the chicken, and you are on the other, so she's kind of sandwiched between the two, allowing only two sides of the table as escape routes. Or, if other people are observing or helping with training, you can have one person on the opposite side of the table so both of you can "body block" in case your chicken wants to hop off and go exploring. Chickens are smart little prey animals, and they are always looking for escape routes! If your hen starts to look like she's going to jump off the side of the table, just gently put your hand or arm in front of her, at eye level, so she doesn't. If you put your hand too low, she'll just hop over it. Put your hand at her eye level, and she'll probably just walk the other way.

Once your chicken starts to understand that the table is where all of the action happens and where she gets all of the treats, she'll probably want to stay on the table a little bit longer each time. When she's pretty good about staying on the table, gradually move the table away from the wall and more into the center of the room.

Wearing jewelry: Since chickens' "bird brains" tell them to peck at anything shiny ("Oh, look! Something shiny!"), it's a good idea to take off all jewelry when you're doing chicken training. It's not a bad thing, per se, to wear jewelry when you're training your chickens, but not wearing it will help prevent them from being distracted.

Distractions: When you're first training your chickens, make sure that you have almost no distractions. Pick a time when the house is quiet, the other animals are put away, and maybe the kids are at school, to bring a chicken out for her first couple of training sessions. The fewer distractions, the better, because chickens are extremely easily distracted. Start to add in low-level distractions when you've had your chicken on the table for a few training sessions and you notice that she's been doing well (i.e., she's taking the treats from you and understanding that the movement is what is getting her the treats). A low-

"Oh, look! Something shiny!" Shiny items, such as jewelry, are sure to distract your chicken.

level distraction could be another person sitting in the room very quietly or an open window that lets her hear some of the outside noises. You don't want to overwhelm your chicken by allowing too many distractions too soon, or else she is just simply not going to be able to pay attention and learn.

Bad time of day: There are good times of day and bad times of day to train your chickens. Once you have training established, it will be easier, but you have to remember that they do have limitations. I love training chicks that are younger than four months old because it's pretty easy to train them at any point during the day. Thinking about how they are most active (and hungry!) early in the morning, this makes a great training time. Oftentimes, in the afternoon, they are going to take their dust baths or relax. This may not be the best time for training. After they are a bit older and have started laying eggs, you want to watch your bird to see what time she usually lays. I can guarantee she's not really going to be thinking about chicken training, she's going to be thinking about laying her egg! My girls are also pretty religious about when it's time to perch. If you try to ask me to think and problem solve when it's past my bedtime, I'll probably get really grumpy and snap at you! So, even though you may be ready to have fun at 9:00 p.m., it's probably not the best time to train your chicken.

CLICKER TRAINING TERMS

I must extend a a huge thank you to Karen Pryor for permission to use the following terms from her Glossary of Terms for Clicker Training (visit *www.clickertraining.com/glossary* for the full glossary) in this book. As I've mentioned, Karen is a founder and proponent of clicker training, and she's also the author of some of my very favorite dog training and behavior books.

Here are some terms that I talk about throughout this book. I've used Karen's definitions here and have added, in some cases, my own comments and substituted examples found in dog training with examples found in chicken training; my additions appear in *italics*. The terms and methods listed are not limited to clicker training and certainly not limited to chicken training. (Here's a secret: I've used some of these methods on my children! *Shhh...*)

Aversive: Any circumstance or event that causes pain, fear, or emotional discomfort. *We will not use aversives in our training, and I hope that you don't use them at home.*

Back-chaining: Training the last behavior in a chain first, then training the next-to-last behavior, then the behavior before that, and so on. Back-chaining takes advantage of the Premack Principle (see page 77). Example: I want my chicken to walk along a ladder and, at the end of the ladder, ring a bell. I would start training her to ring the bell first.

Behavior: Any observable action that an animal does. This can even mean standing still and then turning the head to look at something.

Behavior chain: A series of behaviors linked together in a continuous sequence by cues and maintained by a reinforcer at the end of the chain. Each cue serves as the marker and the reinforcer for the previous behavior, and the cue for the next behavior. *Example: I want my chicken to walk along a ladder and ring a bell at the end of the ladder. I may click for every couple of steps she takes and then click when she reaches the end of the ladder, but she gets her treat only after she has rung the bell.*

Bridging stimulus: An event marker that identifies the desired response and "bridges" the time between the response and the delivery of the primary reinforcer. The clicker is a bridging stimulus.

Chaining: The process of combining multiple behaviors into a continuous sequence linked together by cues and maintained by a reinforcer at the end of the chain. Each cue serves as the marker and the reinforcer for the previous behavior and as the cue for the next behavior.

Classical conditioning: The process of associating a neutral stimulus with an involuntary response until the stimulus elicits the response. *Example: Pavlov's dogs. The neutral stimulus (the bell) elicits the response (salivating).*

Clicker: A toy noisemaker. Animal trainers make use of the clicker as an event marker to mark a desired response. The sound of the clicker is an excellent marker because it is unique, quick, and consistent. *There are several different types of clicker available. I like the old-school type that is fairly inexpensive and looks like a little plastic box. Some clickers allow you to adjust the volume of the click.*

Clicker training: A system of teaching that uses positive reinforcement in combination with an event marker.

Combined training: A type of training that uses all five principles of operant conditioning and a marker signal (clicker) to modify behavior.

Compulsion training: The traditional style of dog training, where the dog is modeled or otherwise compelled to perform the behavior and physically corrected for noncompliance. *This is not at all effective with chickens!*

Conditioned punisher: A conditioned stimulus that signifies an aversive is coming. [It is] used to deter or interrupt behavior; if the behavior halts or changes, the aversive may be avoided. *An example would be noticing that your chicken is about to jump off your training table and quickly throwing your arm in the air to interrupt her from doing so.*

Conditioned reinforcer: A neutral stimulus paired with a primary reinforcer until the neutral stimulus takes on the reinforcing properties of the primary. A clicker, after being repeatedly associated with a food treat or other reinforcer, becomes a conditioned reinforcer.

Conditioned stimulus: Any stimulus that has preceded a particular behavior or event sufficiently often to provoke awareness or response. Clicks and cues are both examples of conditioned stimuli.

Consequence: The result of an action. Consequences frequently, but not always, affect future behavior, making the behavior more or less likely to occur. The five principles of operant conditioning describe the potential results.

Continuous reinforcement: The simplest schedule of reinforcement. Every desired response is reinforced.

Correction: A euphemism for the application of a physical aversive. The aversive is intended to communicate that the animal did something wrong. In some cases, the trainer then guides the animal through the desired behavior. The application of an aversive followed by desired behavior is considered instructive, thus the euphemism "correction."

Counterconditioning: Pairing one stimulus that evokes one response with another that evokes an opposite response, so that the first stimulus comes to evoke the second response. *For example, a chicken may be afraid of a friendly family dog. When the dog is near, the chicken is fed yummy strawberry bits. The goal is to help the bird have a positive association with the dog. You must do this gradually, though, so you don't accidentally cause the chicken to have a fearful association between strawberries and "wolves in the henhouse."*

Criteria: The specific, trainer-defined characteristics of a desired response in a training session. The trainer clicks at the instant the animal achieves each criterion. Criteria can include not only the physical behavior but also elements like latency, duration, and distance.

Cue: A stimulus that elicits a behavior. Cues may be verbal, physical (i.e., a hand signal), or environmental. *For example, when I lift the lid of the food bin, my chickens come running. The cue is the lifting of the lid to cause the chickens to run toward me.*

Desensitization: The process of increasing an animal's tolerance to a particular stimulus by gradually increasing the presence of the stimulus.

Differential reinforcement: Some responses are rewarded and others aren't. *Example: If you want your chicken to peck only at the orange circle, you would not reward her for pecking at other colored circles. Differential reinforcement is not a schedule of reinforcement.*

Environmental reinforcer: Anything in the environment that your animal wants. Trainers can use access to these things as powerful reinforcers for desired behaviors. *Example: Your chicken wants to go outside. You can ask for a behavior and then let your bird's compliance (or noncompliance) determine whether she gets to go outside.*

Event marker: A signal used to mark desired behavior at the instant it occurs. The clicker is an event marker.

Extinction: The weakening of behavior through nonreinforcement or "ignoring" the behavior. In extinction, nothing is added or removed from the environment. For example, a treat lies on the other side of a fence. The chicken tries to reach it but cannot. Because reaching for the treat doesn't work—because it isn't reinforced through success—the bird will eventually quit reaching for the treat.

Extinction burst: A characteristic of extinction. If a previously reinforced behavior is not reinforced, the animal will increase the intensity or frequency of the behavior in an attempt to earn the reinforcement again. If the behavior is not reinforced, it will diminish again after an extinction burst. *Example: There is a treat on the other side of the fence, and your bird was able to reach it. Then you put another treat in close to the same spot, but just scarcely out of reach. Your bird would try very hard to reach the treat before she eventually gave up.*

Fixed interval: A schedule of reinforcement in which the trainer reinforces a desired behavior after a specific period of time—for example, every minute.

Fixed ratio: A schedule of reinforcement in which the trainer reinforces a desired behavior after a specific number of responses. Two-fers and three-fers (*see definitions in this glossary*) are examples of fixed ratios.

Four quadrants of operant conditioning: An incorrect reference to the commonly seen chart illustrating the concepts of reinforcement and punishment. This description is misleading in two ways. It neglects to mention extinction, and it implies that all principles of operant conditioning are of equal value in a training program.

Habituation: The ability to get used to and stop reacting to meaningless stimuli. *This is difficult, but not impossible, for chickens. They are highly instinctual but adaptive creatures.*

Interval reinforcement: The trainer reinforces according to a time schedule. In a fixed interval, the trainer reinforces the desired response after a specific period of time—for example, every minute. In a variable interval, the trainer reinforces after varying periods of time within a certain time frame.

Jackpot: A mega-reward given after a particularly exceptional effort.

Latency: The time between the cue and the response. Initially, that time is zero—or as close to immediate as possible. *In other words, when you first start training, make sure that the treat comes immediately after you click.*

Luring: A hands-off method of guiding the animal through a behavior. *For example, you can use a food lure to guide a chicken into a coop.* This is a common method of getting more complex behaviors. Lures are usually food, but they may also be target sticks or anything else the animal will follow. *I like to use a simple wooden kitchen spoon and then transition to just my finger pointing.* Trainers must be sure to fade the lure early.

Marker: A signal used to mark desired behavior at the instant it occurs. The clicker is a marker.

Modeling: A technique used in traditional training to get behavior. At the outset, the animal is physically guided, or otherwise compelled, into doing the behavior. Pushing a dog's rear into a sit is modeling. Clicker trainers don't use modeling because we want our animals to be active participants in the training process, using their own brains to figure out what will earn them clicks.

Negative punishment (P-): Taking away something that the animal will work for to suppress (lessen the frequency of) a behavior. *Example: A chicken may get in your personal space to try to get the treat. By turning away or leaving, you apply P- by removing the treat as well as the attention that she wants.*

Negative reinforcement (R-): Removing something the animal will work to avoid to strengthen (increase the frequency of) a behavior. *Example: You have horses in your barn with your chickens, and you have a chicken that has been stepped on or kicked by a horse when she was nearby. She will start staying away from the horses' area increasingly so that she will be in a "safe area"— the threat of being stepped on is removed by staying in that safe area.* The key to R- is that an aversive must first be applied or threatened in order for it to be removed.

No-reward marker: Intended to be a signal to say, "No, that isn't what I want. Try again." From the operant conditioning perspective, it's intended to add a verbal cue to extinction. However, once something has been added to the situation, it's impossible to know whether a change occurred through extinction or punishment. No-reward markers usually represent an unnecessary level of complexity in a training program.

Operant conditioning: The process of changing an animal's response to a certain stimulus by manipulating the consequences that immediately follow the response. The five principles of operant conditioning (*see page 81*) were developed by B. F. Skinner. Clicker training is a subset of operant conditioning that uses only positive reinforcement, extinction, and, to a lesser extent, negative punishment.

Permanent criteria: Criteria that are found in the final behavior. Permanent criteria should be trained to a higher level of reliability than temporary criteria.

Poison(ed): No longer reinforcing for the animal.

Positive punishment (P+): Adding something the animal will work to avoid to suppress (lessen the frequency of) a behavior. Example: If I want my chickens to stay in the barn, even when the door is open, I may stand outside the barn and yell, clap my hands, or stomp my feet loudly whenever they approach the open door.

Positive reinforcement (R+): Adding something the animal will work for to strengthen (increase the frequency of) a behavior. *Example: Giving the bird a treat for coming near you will increase the probability that she will come near you again.*

Premack Principle: A theory stating that a stronger response or a preferred response will reinforce a weaker response. During that time, psychologists were focused on behaviorism and were studying instrumental conditioning, which involved attempting to change behavior by tying it to certain consequences. Example: Rats might be taught to press a lever for a food reward or to avoid a lever by getting a shock when they press it.

Primary reinforcer: A reinforcer that the animal is born needing. Food, water, and sex are primary reinforcers.

Proofing: Teaching your animal to perform a behavior in the presence of distractions. *This is not an easy task for chicken trainers, but it is possible!*

Punishment: In operant conditioning, a consequence to a behavior in which something is added to or removed from the situation to make the behavior less likely to occur in the future.

Rate of reinforcement: The number of reinforcers given for desired responses in a specific period of time. A high rate of reinforcement is critical to training success.

Ratio: A schedule of reinforcement in which the trainer reinforces desired behavior based on the number of responses. In a fixed ratio, the trainer reinforces the first "correct" response after a specific number of correct responses. "Two-fers" and "three-fers" (see definitions that follow) are examples of fixed ratios. In a variable ratio reinforcement schedule, the trainer reinforces the first correct response after varying numbers of correct responses.

Reinforcement: In operant conditioning, a consequence to a behavior in which something is added to or removed from the situation to make the behavior more likely to occur in the future.

Reinforcer: Anything the animal will work to obtain.

Release word: A word that signals the end of a behavior. After a behavior is strong and on cue, clicker trainers replace the clicker with a release word. *I haven't used a release word with my chickens thus far, but I think they would understand that it meant the end of the training session because I was no longer handing out treats.*

Respondent conditioning: The process of associating a neutral stimulus with an involuntary response until the stimulus elicits the response. The famous example was the discovery by Ivan Pavlov: dogs drooled when they heard a bell that was previously paired with food. Also called classical conditioning.

Secondary reinforcer: A conditioned reinforcer; in other words, a reinforcer that the animal is not born needing. Secondary reinforcers may be as powerful, or even more powerful, than a primary reinforcer.

Shaping: Building new behavior by selectively reinforcing variations in existing behavior, during the action rather than after completion, to increase or strengthen the behavior in a specific manner or direction. *We will use the shaping method a lot in training our chickens.*

Spontaneous recovery: A characteristic of extinction in which a behavior that was thought to be extinct unexpectedly reappears. If the trainer ensures that the behavior is not reinforced, it will disappear again quickly.

Stimulus: A change in the environment. If the stimulus has no effect on the animal, it is a *neutral* stimulus. (*Example: My chickens live with my pot-bellied pigs. When the pigs walk around, the chickens largely ignore them*.) A stimulus that stands out in the environment, that the animal notices more than other environmental stimuli, is a *salient* stimulus. (*Example: A chicken sees a hawk overhead, so it runs under a bush*.) A stimulus that causes a change of state in the animal (i.e., that causes him to perform a specific behavior) is a *discriminative* stimulus.

Stimulus control: A conditioned stimulus becomes a discriminative stimulus (or cue) when it is followed by a specific learned behavior or reaction. The response is said to be "under stimulus control" when presentation of the particular stimulus fulfills these four conditions: the behavior is always offered when that cue is presented; the behavior is not offered in the absence of that cue; the behavior is not offered in response to some other cue; and no other behavior occurs in response to that cue.

Successive approximation: Increasing or altering a behavior incrementally by repeatedly changing the environment to amplify or extend the behavior. For example, increasing the weight of a load or the height of a jump by small increments to amplify the animal's effort to pull a load or jump an obstacle.

Target: Something the animal is taught to touch with some part of his body. A target is generally stationary.

Target stick: A mobile target the animal is taught to follow. Target sticks are often used as lures to shape behavior. *I like to use wooden kitchen spoons as targets for my chickens.*

Temporary criteria: Criteria that are stepping stones to a final behavior that won't, in their current form, be present in the final behavior. Temporary criteria should be trained only to about 80 percent reliability before "making it harder." If a temporary criterion is reinforced for too long, the animal may be reluctant to change its behavior. *We will use temporary criteria often in our chicken training.*

Three-fer: A training session in which the animal has to perform three behaviors in order to earn one click and one treat.

Timing: The timing of the clicker. Ideally, the click should occur at exactly the same instant the target criterion is achieved. Timing is a mechanical skill and requires practice. The trainer must be able to recognize the behaviors that precede the target behavior in order to click at the very moment that the target behavior occurs.

Traditional training: Compulsion training. Traditional training is characterized by modeling or luring to get the behavior, and the use of negative reinforcement and positive punishment to proof it.

Training period: A preset period of time set aside for training. A training period may be composed of multiple training sessions.

Training session: Either a preset period of time or a preset number of repetitions. Criteria should remain constant during a single session. At the end of a training session, the trainer evaluates the animal's progress and decides whether to make the next session harder or stay at the same criteria.

Two-fer: A training session in which the animal has to perform two behaviors in order to earn one click and one treat.

Variable interval: A schedule of reinforcement in which the trainer reinforces desired behavior after varying periods of time within a certain time frame.

Variable ratio: A schedule of reinforcement in which the trainer reinforces desired behavior after varying numbers of correct responses.

Variable schedule of reinforcement: Technically, this term could mean either a variable interval or a variable ratio. However, most trainers use it to mean a variable ratio.

ADDITIONAL TERMS

Free shaping: A method that requires you to wait for the bird to elicit the desired behavior on her own, and then you reward the behavior to capture the small steps that lead toward the end goal. This is the type of training that we will use most with our birds. I like it because the chickens are fully in charge of the behaviors they are offering, so they will typically learn faster and retain the information better.

Intermittent reinforcement: The practice of rewarding an animal at varying intervals when it does a behavior—not every single time, not every other time, but randomly. I like to do this with my dogs when I'm calling them to come. Sometimes they get a reward, and sometimes they don't. We live out in the country, so I need my dogs to have a very reliable recall. There may be a skunk—or who knows what—nearby. So, when I call my dogs, they need to stop what they're doing (even if it's super-distracting) and come-a-runnin'!

Intermittent reinforcement is pretty powerful. Think of a soda machine. If you put money into a soda machine, you expect a soda to come out, right? You've been conditioned that every time you do this behavior, you'll receive your reward. The reward is a constant, and it's expected. If I treated my chicken every single time she pecked at the dot, she would expect it. But, if one day, all of a sudden, I didn't treat her for pecking at the dot, she'd probably wonder what was going on. If I put money into a soda machine and didn't get my soda, I'd probably figure the machine was broken and I'd walk away, thirsty.

Conversely, if I'm in Las Vegas, gambling at a slot machine, I already know that I'm not going to get a reward each time I put money in. So I'm going to sit there and continue to put money in the slot machine, thinking that maybe I'll get the jackpot this time! Maybe I'll get the jackpot *this* time! Maybe I'll get the jackpot THIS time! I know there is going to be a jackpot at some point, so I keep trying. It's the same thing with animals when you provide them intermittent reinforcement. They know that at some point they will be getting a reward, so they try harder to get it.

Prompted shaping: A training process that uses a food lure or target to elicit desired behavior.

THE FIVE PRINCIPLES OF OPERANT CONDITIONING

The five principles of operant conditioning, developed by B.F. Skinner and all defined within Karen Pryor's list of terms, are as follows:

- Positive reinforcement (R+)
- Negative reinforcement (R-)
- Positive punishment (P+)
- Negative punishment (P-)
- Extinction

Chapter 5

Preparing Your Chicken for Training

W hen I'm working with my chickens, I like to crate them. I think it helps to have clear boundaries, and they seem much more willing to work with me. I equate it to the same mentality of a rough-board horse versus a horse that is in a stall, or a dog that has free rein to go in and out of the house through a doggy door versus a dog that is in a crate while the owners are gone. You're defining boundaries, and you are also creating an environment in which your animal has to look to you for its needs.

When you have chickens that are free to roam about the barn and yard all day long, what do they really need you for? They don't necessarily associate the food in the container with you because it comes from the container, after all! They can do as they please all day and night; in their minds, they don't need a human for anything.

However, if you crate your chickens, either in the house, in the garage, or even in your barn or coop, they now need you for something. They need you to let them out, to feed them, and to give them attention. If you really want to explore the extent of how much you can train your chickens, I recommend crating them or at least giving them clear boundaries.

I crate all of my dogs, and I can tell you that they appreciate having their own areas. Oftentimes, when they are tired, or if there is a lot going on, they will go into their crates on their own. The crates are like their own little bedrooms, and it's the same thing for chickens. They like to feel safe. I use shaping to teach them to go into their crates on their own. Once they have an understanding of going into their crates, I will add a visual cue, usually pointing my hand to the crate, and they will go right in. Actually, even just opening the crate door will cause my chickens to run into their crates. They do like their "bedrooms."

When I'm working with my chickens on a training goal, like getting them ready to go to a seminar, I will crate them in my family room. I find it easier for training because I like to train multiple times a day for just a

Wilderness feels comfortable, not confined, in her crate.

few minutes at a time, so I am putting the chickens in their crates and bringing them out many times throughout the day.

Crating is helpful for transporting chickens, and I believe it helps them to be less stressed. My chickens travel with me to seminars, workshops, and even television appearances. If you are taking your chicken somewhere with you, having her feel comfortable during transport is something you'll want to work on for sure. If your chicken isn't used to going places, she may have already hit her threshold by the time you arrive at your destination and thus may not work for you. Before I need to go somewhere with my chickens, I proof them on crating and car rides. I also start taking them different places simply to get them used to being somewhere other than our home and backyard.

WHAT TYPE OF CRATE?

I like to use plastic crates for my chickens; I think this type of crate makes them feel safe. If you have a wire crate, you may want to put a towel over part of it so the chicken feels safer when going inside it. The crate should be large enough for the chicken to stand up, turn around, and lie down comfortably. It really does not need to be larger than that for training.

Acclimating Your Bird to the Crate

Your training crate can double as your travel crate, reducing the chicken's stress when you need to take her somewhere.

One of my all-time favorite chickens was a Blue Andalusian chicken named Sam. Sam was crate-trained and potty-trained—yes, potty-trained! I would like to take credit for her potty training, but this is something she did all on her own. Sam would not go to the bathroom in her crate. It could be day or night, and she would hold it. Just like a broody chicken lying on her eggs, she would hold it for hours. Then, within about two minutes of being let out of the crate, she would poop. She became wonderfully predictable. I would let her out of the crate and take her straight outside. She would do her business, and then I would bring her back into the house to train. Brilliant! I have to admit, it was awesome. Plus, I didn't have to clean out her bedding each morning.

Not all chickens will become potty-trained like Sam, but many will. The more time you spend with your chicken, the more joy and satisfaction you'll get out of working with her. If you can start crate training your chicken early on, that is ideal. By "early on," I mean around one week

CREATING BOUNDARIES

If you don't want your chicken to be in a crate for hours on end, you can create boundaries in other ways. As long as your chicken doesn't have free range, you will get similar results in respect to how she interprets her relationship with you. I have seen people create a living area for their chicken in their bathroom; they fit the tub to have a nesting area and an area where the chicken can walk away from the nest to eat and go to the bathroom. You can create a similar area of confinement with wire exercise pens (also called "ex-pens") made for dogs or even the plastic ones that are used for toddlers. Just make enough room for your chicken to be comfortable without allowing her to come and go freely. If you live in the city and are keeping your chicken as a pet, this is a great idea because she'll have her own little space.

old; I don't like to start training itty-bitty baby chicks until that age. Plus, that is when they really start eating food, and you can use food as a reward.

Most of what I do with crate training is luring. When I'm working with a chick on helping her feel comfortable in the crate, I have my treats in a measuring cup and offer them to my chick. Once my chick is willingly taking treats from the cup, I slowly move the cup toward, and eventually into, the crate. Once my chick has followed the cup into the crate, I leave the cup in the crate for a few moments and let the chick peck away, and then I will take the cup out of the crate. The first few

Sam was always a very proud chicken.

times, the chick will follow the cup out of the crate again, so I start the process again. Each time, I require the chick to be closer to the crate before I give her a treat. In the beginning, I will give a treat for just being near the crate. But after I've lured the chick into the crate a few times, I start to let her have treats only when she is at the crate door opening, and eventually only when she has followed the treat container all the way into the crate.

Your goal is to help your chick understand that she needs to be in the crate in order to get

the treats. So now that you have your chicken expecting food from your container, you can start to bring it onto a table. The difference between the crate and the table is that when your chicken is in the crate, she is going to feel pretty safe. When your chicken is on a table, she is out in the open. She is exposed to predators. It's going to be much more distracting for her to take food from the container. But, remember, if you want to train your chicken to do tricks or behaviors, you have control of that valued resource. So, think to yourself, *Gee, my chicken hasn't eaten in about twenty minutes, so she's probably getting pretty hungry.* That would be the perfect time to do some training.

From my experience, I've learned that you can take a relatively feral chicken and, within a couple of days, have this chicken learning behaviors and expecting positive reinforcement from you. Chickens are extremely fast learners. After you've done it a couple of times, they've got it.

Step-by-Step Crate Training

Step 1

Have your table ready; waist height is perfect. Place the table next to the wall or ask another person to stand on the other side of the table to prevent the chicken from falling or flying off the table. Place your crate and your chicken on the table.

Step 2

Offer the chicken treats from your measuring cup. Once she is pecking at them willingly, slowly move the cup (lure) toward the crate and eventually into the crate. Once the chicken is in the crate, let her have a few extra pecks of food.

Your chicken needs to feel comfortable on the training table in order to learn and ignore distractions.

LURING

Luring is different from clicking and treating. Luring is more like telling the chicken, "Follow this treat container, and when you get close to it, I will let you eat from it." Luring is just one of the ways you can train a chicken.

Step 3

With the chicken and the crate on the table, lure the chicken into the crate with the treat container as before and let her have a few pecks of food.

Step 4

Again with the chicken and the crate on the table, immediately place the treat container near the crate so the chicken has to move toward the treat container instead of you gradually luring her there. If this doesn't work, show her the treat container, let her have one peck, and then move it all the way to the crate door, which should be only about 10 to 12 inches away from her if she's an adolescent or adult. If she's just a baby, this may be too far away, so place her a little closer to the crate. Remember, you want her to start moving toward the treat container on her own versus you having to lure her all the way.

Step 5

This time, place the chicken and the crate at opposite ends of the table. Place your treat container a couple of inches inside the crate so the chicken now has to walk the length of the table (2 to 3 feet) and go to the opening of the crate door to get the treat. Once her head is inside and she is taking a few pecks of food, lure her around so she is now inside the crate, facing the crate door. At this point, give her a few pecks of her food—she's a clever chicken, after all!

Step 6

Place the chicken and the crate again at opposite ends of the table, with the treat container inside the crate. Your chicken should walk right over to the crate pretty quickly and go over the threshold of the crate door to get the treats. However, don't let her have a treat this time; instead, lure her all the way into the crate so she is fully inside and now facing the doorway of the crate, and then let her have a few pecks.

You may have to repeat this step a few times, but your chicken will quickly learn that when you put her on the table and her crate is present, she should walk over to her crate, go inside her crate, and then turn around in the crate so she is facing the door. At that point, very quickly, bring your treat container to her beak so she can peck a few tidbits.

The foregoing steps teach your chicken to go into the crate. Next, we will work on teaching her to stay in her crate. This next step is easy if you have very quick timing.

Step 7

Place the chicken and the crate at opposite ends of the table. Your chicken should promptly walk over to the crate, go inside the crate, turn around in the crate, face the crate door, and expect a treat to appear. As soon as she does that, quickly give her a treat and then pull the treat container back toward you and then immediately bring it back to her. So, you're treating and then, in one fluid movement, withdrawing the treats and offering them to her again. You want to be super-fast with this movement because you want to bring the treats back to her before she flinches and thinks about stepping outside of the crate. Your bird is amazingly intelligent, and if she thinks that standing still inside her crate (even for a split second) is *not* getting her a treat, then she is going to try something else. You want her to understand that by standing there and staying in the crate, she is going to get more treats.

Repeat this exercise, in which you are treating her, pulling the treats back to you, offering the treats to her, and then pulling them back to you again, about six to ten times, very quickly. Then take a break! I keep my training sessions from two to five minutes each (or shorter with young chicks). Much longer than that, and she will lose interest (and hunger!). Plus, I like to end with my chicken wanting more, so when we start another training session, she will remember the exercise as being fun and rewarding for her.

Remember, the goal of offering her the treat and then quickly withdrawing the treat container when

she is standing in the crate is to gradually teach her to stand in the crate for longer amounts of time without treating her. So, it should go like this:

- Treat—withdraw—treat—withdraw after half a second
- Treat—withdraw—treat—withdraw after one second
- Treat—withdraw—treat—withdraw after two seconds
- Treat—withdraw—treat—withdraw after three seconds
- Treat—withdraw—treat—withdraw after four seconds
- Treat—withdraw—treat—withdraw after five seconds
- Treat—withdraw—treat—withdraw after six seconds

And so on... Eventually, your chicken will learn to walk over to her crate when it is presented to her, go inside, and patiently wait for a treat.

A chicken typically will be very happy in her crate once you shut the door; the crate seems to have a calming effect. The only chickens who mind it are some of my dominant hens and my roosters—probably because they know that they should be out there, doing their jobs! But, for the most part, as long as you are fair and don't leave them crated for too long, chickens seem to enjoy the peace and comfort of sitting around and doing nothing.

Proofing

Once my chickens are comfortable in their crates, I will start proofing them for situations that will arise, such as:

- Lights turning on and off (which is not natural for them in their outdoor barn or coop living environment).
- Household noises, such as doors, people's voices, footsteps on hard floors, doorbells, and the like.
- Being lifted and carried in the crate for short distances. The crate can be a bit wobbly and, in the beginning, may be scary for chickens. They can get used to it—I've even had chickens fall asleep while being carried in their crates.
- Temperature changes, such as when going from the house to the garage, or from

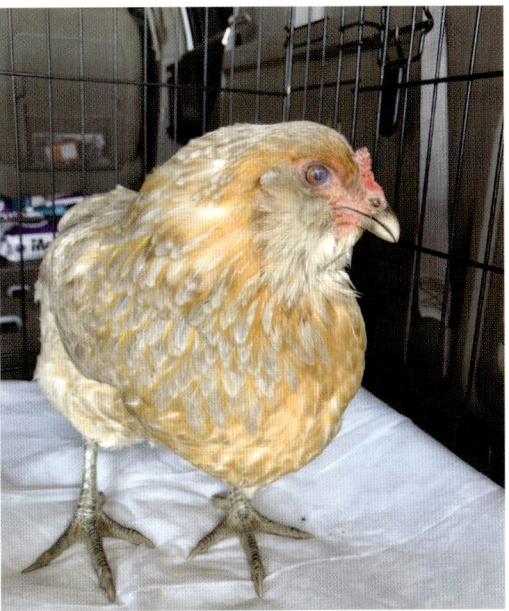

Using the same crate for training and for car rides helps your chicken feel comfortable with both.

BACK TO THE FLOCK

If you take a chicken indoors for training for a few days, where she will be in and out of the crate, be prepared for all that may happen when you put her back with the flock. If you "pluck" a chicken from her flock, she may have to establish her place in the pecking order again when she returns. The flock dynamics may have changed without her, and she may seem a bit like an outcast when she returns. From my experience, this normally lasts less than a day, and then all of the chickens are back to where they were. Chickens have a social hierarchy, and they also have close friends and relationships. Being someone that loves to observe animal behavior, I am very entertained by watching my chickens. Sometimes they can be very sweet, and sometimes they can be downright jerks!

the car to a building. **Note:** Be considerate of your chickens and don't expose them to any temperature changes that may put them at risk. For example, I wouldn't take my chickens out of the barn in the middle of winter and bring them into the house for a couple of days because they will get much too hot, and then it may be dangerous to put them back into the cold barn. The same applies to air conditioning during the summer. Keep this in mind and do what's best for your chickens.

- Eating and drinking inside the crate. I like to use the glass measuring cups for their food and water because they hook nicely onto the crate door and are difficult to spill.
- Your hands reaching into the crate. As much as you work with your chicken on going in and coming out of the crate, there may be times when you have to reach into the crate and gently bring her out. She will be much less stressed if this is something on which you have gently proofed her.
- Hearing and seeing people other than you, or even hearing you talk to people. It's silly to think that hearing me talk is something that my chickens may not be used to, but I don't talk much when I train them.
- A friendly dog. If you have a friendly, calm dog, let him sniff the chicken's crate. However, I wouldn't let an unfamiliar dog approach my chicken's crate because I don't know what he will do, and I want my chicken to feel safe and confident that I won't let anything hurt her. **Note:** Chickens have a pretty cool ability to tune out distractions during training; however, you're not going to want to train your chicken while your dog is walking around underneath the training table unless your chicken is very used to and comfortable with your dog being in the room while you're training.

OFFERED ATTENTION

Whenever I start a training session, I like to first have my chicken's attention. This will help her understand that we are going to be working. I want my chicken to think that I'm the most exciting thing in her environment.

To do this, I am going to start rewarding *offered attention*. There is no way I can force my chicken to pay attention to me, but I can reinforce her when she offers attention. This may start off as a glance in my direction and work its way up to her watching me and starting to offer additional behaviors to get the treat. It's also a good warm-up to training and letting your girl know that you have yummy treats waiting for her.

Have your bird on the table and your treats ready. You'll have to be quick! Click-treat anytime your chicken even glances in your direction. If she catches on very quickly, she may just stop and look up at you—click-treat. If you're thinking, "My chicken just keeps staring at me," then take just one or two steps to the side of the table so she has to turn her head or body to continue to look at you. When she gets really good at this, start to provide mild distractions, such as moving one of your hands to the side of the table and then waiting a second for her to look back at you. The split second she does, click-treat. You can gradually make your distractions more difficult, such as tapping on the table with your empty hand (the hand that is not holding the treats). Tap just once—you want to make it mildly distracting, not overwhelming, for her. Wait for her to look back at you, and then click-treat. Try having someone stand still on the other side of the table; sometimes a person simply standing there is a distraction. If your chicken gets good with a person standing still, progress on to having that person make subtle hand gestures or movements.

If you provide a distraction and your bird completely loses focus, you'll need to go backward in your training a bit. Go back to a milder distraction, where she still understood that she was supposed to look at you. Gradually work your way up to the more difficult distraction again. The goal is not to completely distract your bird but to let her know that even when things are going on around her, it's good for her to pay attention to you.

Work on offered attention along with the other behaviors that you are teaching your chicken. Once she is pretty good at offering attention, start off each training session with just a few moments of practicing it.

Chapter 6

Chicken Tricks

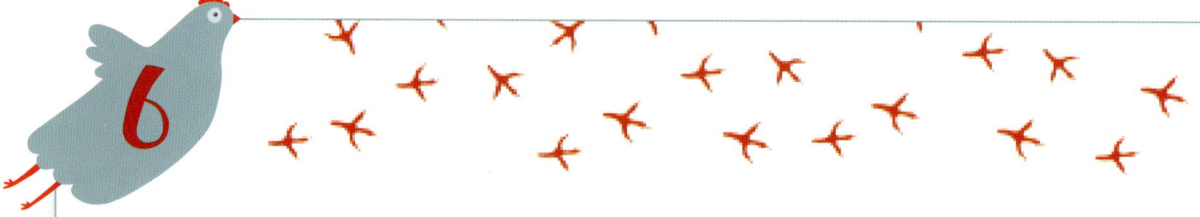

ow you're ready to start training tricks! You've conditioned your chicken to the clicker. She is used to being on a table, and you've even done a little work with offered attention. So let's get started with the fun part!

Trick: Peck at the Dot

This is the first trick I like to teach. Essentially, what you want to do is take a natural behavior—something that your chicken is already doing naturally—and put it on cue, meaning that you train her to do it on command. Pecking is a natural behavior for chickens. They will peck at just about anything—bugs, crumbs, specks, the freckles on my hand—to see if it's edible, so this is a great one to start with.

The behavior we want to teach is to peck at a dot on a paper plate. What you're going to do is make a circle, about a centimeter in diameter, in the center of a small paper plate.

To start, place your chicken on the table. Remember, she's already been on the table a couple of times, so she's comfortable being there now. You've also already done some clicker training with her, so you are setting your chicken up for a successful training session.

With your chicken on the table, take the paper plate and gently slide it near her. Your chicken is very curious, so she's probably going to look at the paper plate. The split second she even looks at the paper plate, you're going to click and offer her a treat. You're *shaping* the behavior that you want. Think about what your chicken needs to do to peck at the dot and the steps you need to take to help her figure out that pecking at the dot is what you are looking for. Your chicken isn't going to automatically know that you want her to peck at the dot. You need to shape the behavior by rewarding her for baby steps that will lead her to the behavior you are looking for. Think about the first thing the chicken will need to do to eventually peck at the dot. Yes! She'll need to *look* at the dot first.

The split second your chicken looks at the dot, click and treat. She might look at the dot again, in which case you click-treat again. Do this with her three or four times in a row. At that point, your chicken is probably starting to understand that if she looks at the dot, she's going to get the click (which means that the reward is coming) and then her reward.

With shaping, or operant conditioning, make sure that the animal is doing the behavior about 80 percent of the time before you move on. If my chicken is doing the behavior about 80 percent of the time, I can safely assume that she understands why she is getting rewarded. However, if I were to always click and treat my chicken for looking at the dot, I'm not going to get any further along in her training. So once she's doing the behavior about 80 percent of the time, I'm going to stop clicking and rewarding for that behavior. I know that sounds a little strange, but there is a reason behind it: if your chicken thinks that she gets a click and reward every time she looks at the dot, but then the reward never comes, she is going to do something else. She may look at you. She may walk away from the paper plate. She may stand there and poop! Or, she may look at the dot and then reach her head down toward the dot. *Bam*! That is when you click-treat again.

Your timing skills are going to really come into play because chickens can move their little heads so quickly, and you want to make sure that your timing is just right. You want to click-treat right as she's moving her head down toward the dot. You want to be able to click her before her head is all the way down and before her beak actually touches the dot—you want to capture the behavior right as it's happening, not right after she's done it. For example, if you try to time your click right as she's pecking at the dot, your click may come a millisecond after she's pecked, so she may think that she's getting clicked/rewarded for starting to lift her head away from the dot. Yes, chickens are that fast!

The first phase of the behavior of pecking at the dot was simply looking at the dot. You clicked and rewarded your chicken. Then, the second phase is the chicken moving her head down toward the dot, for which you also clicked and rewarded your chicken. Once your chicken is moving her head down toward the dot two or three times in a row, that's when you stop clicking and rewarding that behavior. Sound familiar?

Your chicken will probably stop and look up at you as if to say, "What the heck?" This is the fun part about shaping because you are giving your animal time to think about what it should do next. Now, because you're training a chicken, this might be a perfect time to end this training session.

When I train a chicken, I love to do a little bit of training and get her to the point of "Oh! I think I get it!" At that point, stop. Put the chicken back in her crate and let her chill out for a little while. When you bring her out again, twenty or so minutes later, to start your training session again, your chicken will be thinking, "I remember this!" She will be that much more excited for training time. So if you've gotten your chicken to do a couple of steps in your shaping process of pecking at the dot, it's a perfect time to put her in her crate and let her think about what you were just working on. Then, when you bring her out again and show her the paper plate with a dot on it, your training session may start from square one, but it should progress much faster the second time, and even faster the third time. So, if your last training session consisted of your chicken *almost* pecking at the dot, you should be able to get her to *actually* peck at the dot in the first few moments of your next session. Then, in subsequent sessions, she should immediately go to the dot and peck at it because she's starting to understand that pecking the dot is what is enabling her to get treats.

After your chicken has pecked at the dot a few times in a row, you can start to proof her. At this point, everything happens very quickly. I tell you—once they get it, they've *got* it! Now, instead of keeping the plate in just one place, you can start moving it around. Let your chicken peck at the plate and then gently (yet quickly) slide it a couple of inches to the left. Let her peck at it. Slide it a couple of inches to the right. Let her peck at it. After that, start to move the plate over to one side of the table so your chicken has to turn her body and maybe walk a step or two to the plate to peck at it. Start picking the plate up so it's no longer flat on the table. Hold the plate in your hand and turn it vertically instead of horizontally. Hold the plate up so it's at eye level with your chicken. Once your chicken starts getting really good at pecking the dot, you can hold the plate up even more so she has to jump up a little bit to peck at the dot.

Look at that—you did it! Now you and your chicken are ready for target training.

Trick: Target Training

In my opinion, target training is one of the most useful behaviors that you can teach your chicken. If your chicken is target-trained, you can essentially lead her anywhere you want her to go. You can take her on walks, have her go into her coop when it's not necessarily time to roost, lead her into her crate, do little chicken agility or obstacle courses—the behaviors you can have your chicken offer with target training are almost endless.

To give you an example of the usefulness of target training, I want to share a story about a pig named Winnie. When I was on the board of directors of Heartland Farm Sanctuary, Heartland had the opportunity to rescue a little piglet. This piglet had fallen off a transport truck, and a Good Samaritan had found her and contacted Heartland. Winnie was pretty beat up and had a very sad story. Stephanie, one of my good friends and a longtime Heartland volunteer, contacted me about Winnie. She asked if I could come out to Heartland to do some training with her. Of course, I was thrilled and excited about the opportunity to train this baby pig. This was before I had my own pot-bellied pigs, so I hadn't trained a pig before. The Heartland staff had high hopes for Winnie to be one of the ambassadors for their sanctuary, and they knew that, therefore, she would have a very social life ahead of her. They wanted her to be prepared for plenty of positive interactions with people.

This little piglet was just the size of a puppy when we first started working with her, but we knew that in a very short time, she was going to be a 500-pound (at least!) farm pig. And 500-pound farm pigs pretty much think about eating all the time! If they're not eating, they're sleeping or rooting—but most of the time, they prefer to be eating. The main behavior we wanted to work on with Winnie was to make sure that she had good piggy manners. A lot

Training baby Winnie (left) versus training Winnie at five months old (right).

of volunteers would be coming in and out of her stall, and many of the volunteers would be children. We knew that Winnie would quickly understand that when her stall door opened up, food was coming in. Pigs are, well, pigs! We wanted to make sure that she didn't knock over people who came into her stall or, even worse, run them over in an attempt to get out. The food preparation table was, coincidentally, right across the aisle from her stall.

We figured we'd need Winnie to have a little bit of impulse control so she could stay in her stall and allow people to open the door safely, walk into the stall, and then pet her, feed her, examine her, or even give her a massage.

But could we actually teach impulse control to a pig? For a pig, impulse control is a very tough behavior to learn. We figured that it would be difficult to teach her a "wait" or "stay" command, so we decided to help Winnie stay in her stall when people came in by target-training her. Instead of making her stay still, we gave her a job to do. We taught Winnie that the only way she could make someone come into her stall was by walking away from the stall door.

We took a bright orange Frisbee and nailed it, at her snout level, to the opposite side of her stall from the door. We then used shaping to teach Winnie that when somebody came to her stall door, the only way that the person was going to come in was if Winnie touched her target (the frisbee). This allowed the volunteers or veterinarians to come into her stall safely because when they approached the door, Winnie would walk away from her door and go touch her target. People could enter the stall with enough time to close and latch the door safely behind them without letting a massive pig loose. *Ta-da!* Winnie was target-trained.

When I teach target training to my chickens, I like to use a wooden kitchen spoon as my target. It probably would be easier to just use my fingers, but I don't want my chickens pecking at my fingers—that hurts! I also like using the wooden spoon because it's easy to draw a black dot on the wooden spoon just like you did on the paper plates that your chickens have already learned to peck. This is a pretty easy behavior to transfer from the paper plate to the wooden spoon.

To start the target-training session, simply place the wooden spoon on the table with the black dot facing up, toward your chicken. Since your chicken already knows how to peck at the dot on the paper plate, she should probably peck at the dot on the wooden spoon rather quickly. So when she pecks at the dot on the spoon once or twice, and you've clicked and rewarded her each time, pick up the wooden spoon and very slowly start moving it around the table, just a couple of inches at a time. Move it to the left, move it to the right, move it up, move it down. Move it so that your chicken has to reach down and maybe even take a couple steps backward in order to peck at it.

As I mentioned, you can expand on this behavior and teach your chicken agility. You can also teach her to come when called by using target training. If your chicken is at a distance from you, and you show her that target (the dot on the spoon), she is going to come running to you so she can peck it—your chicken has just come to you when you called!

Sprinkles approaches the target and then follows it with her head.

TRANSPORTING YOUR CHICKEN

I can imagine that, if you're an animal that has never been in a car, going on a car ride might seem pretty crazy to you—the ground moving underneath you, turns, bumps in the road, and so forth. Not to mention the radio playing and heat or air conditioning blowing. If you're going to transport your chicken, start with baby steps.

I always transport my chickens in crates. They will be safe, and you won't have to worry about them pooping in your car twenty times during the ride. Here are the steps to get your chicken used to riding in the car:

Day 1: Take your chicken, in her crate, into the car, with the car turned off. Just place her crate in the car with the car door open and sit with her in the car for a few minutes. She may make fussing sounds for a bit. Once she has calmed down, bring her crate out of the car and bring her back into the house (or barn or coop).

Day 2: Bring your chicken, in her crate, into the car. Let her sit with you in the car for a couple minutes. As long as she seems calm, go ahead and start the car. Just have the car running, no radio, no driving. You can turn on the air conditioner or the heat to keep the temperature comfortable. Your chicken will sense the temperature difference, the smell, and the sounds and vibrations of the car. Spend a few minutes with the car on, and then bring her out of the car.

Day 3: Bring your chicken, in her crate (you've probably guessed by now that she'll always be in her crate when you transport her), into the car. Turn the car on so she can experience the temperature, balance, vibrations, and sounds. Turn on the radio and don't forget to talk to your chicken. When I work with my animals, it's usually just me and them, and I find that I don't really talk that much. I'm focused on them, on my timing, and so forth. So I have to remember to talk! I'll be talking when I bring my chickens to seminars and other events, so they will have to get used to hearing me.

Day 4: Bring your chicken (yes, in her crate!) into the car and go for a short drive—maybe just around the block. The movement of the car will be what's different on this day. After you've driven her around a bit, she will probably be pretty accustomed to being in the car. You and your chicken are ready for your first outing!

I've seen zookeepers using target training with their animals. I've watched them train polar bears, seals, and badgers. It was interesting to see the animals being given opportunities to use their minds at the zoo. I also liked how the designated training times brought crowds of people to watch and ask questions. I think it's important for people to understand that the creatures at the zoo are highly intelligent animals that need enrichment on a daily basis.

You can get creative and target your chicken to do almost anything! For example, use the spoon to point at things. Teach your chicken to peck at a small ball and set up little goals as if she were playing soccer. Or target her to peck at a small door so that it opens. You may have to use shaping if you want her to peck hard enough on the door to make it open. Use your target stick to teach your chicken to jump up and land on your arm, a perch in the coop, a swing outside, and so forth.

I've seen people target-train their chickens with laser pointers, and you can, too. You can teach your chicken to peck at the laser-pointer dot. You can train her to follow the dot or peck at objects where she sees the dot, for example, the keys of a toddler-sized toy piano. I've seen videos of chickens playing piano, and it's adorable! You can easily teach your chicken to do this simply by pointing the laser pointer at the keys you want her to peck.

Trick: "Are You a Chicken?"

I think that this is one of the coolest tricks that you can train a chicken to do. It will help your friends truly understand how intelligent chickens are. Part of the reason I have fun training chickens is because I like to educate people who really have no clue about these wonderful little birds. Many people think that chickens are "bird brains," and they see chickens much differently than they see smart dogs or cuddly cats. I think that if people just knew how clever chickens are, they might advocate more for these bright birds.

I love to have my chickens show off the "Are You a Chicken?" trick when they are on TV. I feel that this trick leaves people in amazement. While I would love to take credit for the idea, I found this trick through an amazing organization in Australia called Edgar's Mission. Edgar's Mission had an

Your bird will look like a genius when she can pick out the chicken.

With my thumb ready on the clicker, I click the split second before she pecks the orange wood.

adorable little rescued battery hen (confined hen used for egg-laying), named Little Miss Sunshine, that they rescued in 2013. Little Miss Sunshine passed away in 2017, but her story, along with photos and videos, are still on their website (*www.edgarsmission.org.au/ category/blog/little-miss-sunshine*). She was very endearing and much loved, and she knew her name and she some great tricks. Watching her in action helps people understand that chickens are intelligent, sentient animals.

To teach this trick, I first went to a toy store and bought a wooden puzzle with farm-animal pieces: a cow, a horse, a pig, a duck, a sheep, and, of course, a chicken. The goal is to be able to lay all of the puzzle-piece animals out on the table and then ask your chicken, "Are you a chicken?" And then your chicken pecks at the wooden chicken—brilliant! Your friends will be amazed that your chicken actually knows that she's a chicken!

Because you now know a little about how chickens learn, you can understand the idea that this is all a training exercise. If you think about how the chicken perceives this "trick," she understands that when she is in the presence of the little orange piece of wood, she should peck on the little orange piece of wood—hence, making us silly humans think that she understands that she's a chicken!

The way to train "Are You a Chicken?" is to start with only the chicken puzzle piece present. Start off just like you did when you trained your chicken to peck at the dot on the paper; this time, you'll teach your chicken to peck at the chicken puzzle piece. The moment your chicken even looks at that puzzle piece, quickly click-treat. When she looks at the puzzle piece again, click-treat. When she starts moving her head toward the puzzle piece, click-treat. If your chicken walks away from the puzzle piece, do not click-treat. She doesn't get any type of reward. Don't make any noises or movements because you don't want to unintentionally reward your chicken for doing a behavior that you do not want at that moment. You also don't want to inadvertently guide her into doing what you want her to do. Even a slight hip tilt toward the puzzle piece could "give it away." You want your chicken to figure it out on her own.

You should be able to get your chicken to peck at the puzzle piece very quickly. She has already figured out that when you place something on the table, she pecks at it, right? So now that she's "got it" and has pecked at the puzzle piece consistently a few times, you can add a second puzzle piece. When choosing your second puzzle piece, pick one that is very different visually from the chicken puzzle piece. My chicken puzzle piece is black with red around its head. Its feet and beak are orange, and it has white dots on the side of its body. So the puzzle piece I choose as my second piece is the duck. The duck is all white with a little bit of orange on its beak and feet, so it looks very different from the chicken.

Take the chicken puzzle piece and the duck puzzle piece and present them both to your chicken. If she pecks at the chicken, she's going to get a click and a reward. If she looks at or pecks at the duck, she's going to get nothing from me. I'm going to stand still; I'm not going to move. I'm hardly even going to breathe or blink. But the split second she goes back to looking at or pecking at the chicken, she's immediately going to get a click-treat. After you do this a few times, your chicken is going to start understanding that when in the presence of the chicken puzzle piece, she pecks only at the chicken puzzle piece.

At that point, you can start proofing your chicken by not just placing the two puzzle pieces in the same pattern on the table (for example, chicken on the left, duck on the right). You want your bird to understand that you want her to peck the chicken, not just the object on the left side of the table. Start moving the pieces around just like you did with the paper plate with the dot on it. Move the pieces to the left a little bit. Then move them to the right a little bit. Switch them around so maybe the chicken puzzle piece is closer to your chicken and the duck puzzle piece is farther way. Then, as soon as she pecks at the chicken puzzle piece, move it farther away and the duck puzzle piece closer, and so on.

Once you've proofed your chicken to understand that *in the presence of the chicken puzzle piece, she pecks only at the chicken puzzle piece*, adding in a third, fourth, or fifth puzzle piece is going to be much easier. The hardest step in this process is when you present the second puzzle piece. After that, your chicken should be pretty well proofed. She will probably understand that no matter how many puzzle pieces you present to her, as long as that chicken puzzle piece is on the table, that's the only one she's going to peck at. It's amazing to watch your chicken looking at all of the puzzle pieces on the table. You can just tell that her mind is working.

FEATHERS

One of the first chickens I trained was Feathers, a curious and sweet little Splash Andalusian. She would sometimes look at me when I'd present her with something just a little more difficult than what we had just done. It was amazing to watch. I'd do something like add another puzzle piece to the table or move the pieces around. Or I'd try something with her that she hadn't seen yet. She would stop, look at the object, and then turn her head around and look me straight in the eye as if to say, "Really?" I took her to many seminars, and each time she did this, everyone would laugh and laugh. She was a special chicken to me.

When I'm showing people how smart chickens are, this is one of my favorite behaviors to demonstrate. I also wanted to help my chickens understand that they can do the same behavior in different contexts, and I didn't want my chickens to get bored with the puzzle pieces. For a bit of variation, I took three pieces of foam and, with my black marker, drew the outline of a horse on one piece, the outline of a pig on another, and the outline of a chicken on the third.

Trick: Crate Training

There are many different ways to train a chicken to do a specific behavior. In chapter 5, I spoke about getting your chicken to go into her crate to help her become more comfortable. For this, we used luring. Now we will talk about using target training with the crate. Luring gets the behavior you want with minimal training. So, if you want to get your bird into the crate quickly because you have to go somewhere and you don't want to just grab her and plop her in there, luring is great. Target training, however, will help you solidify the behavior, and your chicken will eventually go into the crate on her own without the target and will stay there on her own.

Chickens don't usually try to get out of their crates like some dogs do. Once they are in their crates, they seem pretty content. I think that being in a smaller covered area probably makes them feel safe. As long as you give them opportunities to come out of their crates and be chickens, they should not be bothered by being crated. But if you leave them in there too long, they're going to go crazy.

What you'll need:

- An inexpensive wooden kitchen spoon
- Black pen or marker to make a round dot, about a centimeter in diameter, on the flat part of the spoon

For crate training, I use a travel crate that allows the chicken to stand up, turn around, and lie down comfortably but does not provide a lot of excess space.

- Crate
- Treat container (such as a measuring cup with a handle)
- Clicker (I like to Velcro my clicker to the handle of the measuring cup so I only have to use one hand. It makes training much easier.)
- Treats—shredded cheese, shredded coconut, cut-up strawberries, ground beef—whatever are your chicken's favorites

When you're training your chicken how to go into the crate, there a few different phases:

- The chicken walks into the crate.
- The chicken turns around and faces the doorway.
- The chicken lies down in the crate.
- The chicken stays in the crate.

Remember, with operant conditioning (which is what you are doing when you target-train), you're going to reward baby steps. I don't want you to think that during your very first training session, even though you've already target-trained your chicken, you're going to be able to train your chicken to walk all the way into the crate. There is the possibility that this may happen—and, if it does, that's awesome! But let's just remember to start with baby steps. Envision your final goal but click-treat the baby steps of your chicken getting close to the crate and eventually going into the crate.

You can almost see the wheels turning as Sprinkles looks at the dot and moves to peck at it.

Place your crate and your chicken on the table. Have your treats and your target stick (wooden spoon) ready. Thinking about what you want your chicken to do, the *very* first step is for her to turn her body toward the crate. She's not going to walk backward into the crate! So, first use your target stick to direct your chicken toward the crate. Even just turning her head toward the target stick earns her a click-treat response. Remember, our chickens are lightning fast, and if we're not giving them correct and quick guidance on what we're expecting of them, they're going to move on and ignore us. Once your chicken is on the table, it's go time!

Hold the target stick (spoon) just a couple of inches away from her head (or if that's too much for her, and she doesn't turn toward it, just one inch). Click-treat as soon as she goes near the target stick or if she pecks at the dot on the target stick. Use the target stick to guide her to turn her body toward the crate. Click-treat just about every step that your chicken takes toward the crate.

You might get into a situation in which once your chicken reaches the threshold of the crate, she stops. If this happens, and it becomes an issue, you might want to regress a little in your training. You can try taking the door off of the crate. You can try taking the top portion of the crate off so that you're targeting your chicken to walk into just the bottom portion of the crate. Whatever you do, be sure to watch your chicken and her behavior. It's also a great idea to keep a training journal of your training. Take short notes after each training session to help you remember what your chicken did and what you observed. If she's feeling a little too nervous about walking into this dark, cave-like, scary thing, make it a little bit easier and a little bit less scary for her.

Assuming that you're going to be able to get your chicken to walk into the crate with the top on and the door on and open, you're going to run into a problem of being able to treat your chicken once you hit the threshold of the crate. You won't be able to easily put that measuring cup directly in front of her beak because the front part of her body is going to be in the crate. There are two things that you can try. We talked about how the click tells the chicken that she did something right, and it also means that a reward is coming. If you click your chicken right as she is stepping across the threshold of the crate, or right as her head goes into the crate, she still is going to understand that that is why she's getting a reward. So you can still click right as she's going in, and then she'll probably turn her head back out of the crate to get the reward. That is totally fine! However, it would be ideal if you could actually get your treat cup into the crate while your chicken's head is going into the crate. This way, you are not having your chicken go into the crate, come out of the crate, go into the crate, go out of the crate, and so on. You want your chicken to understand that this behavior is going to be fluid and that you eventually want her to go all the way into the crate. So, if you can, lean down gently and reach your hand into the crate with your chicken so that you can click and treat her while she continues to walk into the crate.

The process will look like this:

Step 1: Get her to turn her head toward the crate.

Step 2: Get her to walk toward the crate.

Step 3: Get her to walk into the crate, even if it's just one step! When you're successful at step 3, use your jackpot. You want her to understand that what she did was super-amazing. So once you have her stepping into the crate, it is a milestone—jackpot time!

Step 4: Get her to walk all the way into the crate. Once you have her going into the crate, and then going in and turning around, that is another milestone—another jackpot! When she is walking into the crate willingly, it's great to give her a jackpot because it helps prevent her from walking directly out of the crate.

Step 5: Get your chicken to stay in the crate. Chickens like movement, so getting her to walk into the crate might be the easiest part of this trick. Getting her to *stay* in the crate might be a little tougher. This is where your operant conditioning comes in.

Remember, operant conditioning teaches us to reward a behavior until the animal is doing that behavior about 80 percent of the time, at which point you stop rewarding that behavior because you want your animal to move on to the next step of that behavior. So, if I want my bird to stay in the crate, I'm going to reward her for being in the crate, but what happens when I remove the treat container from the crate? I want her to understand that even though the treat container is moving away from her, outside of the crate, I still want her to stay in the crate. Do you think you have good timing? This is a great opportunity to work on it.

You need an extremely quick rate of reinforcement to help your chicken understand that staying in the crate is what you want; you must click and treat her before she even has the chance to walk out of the crate. So, you're going to click (because she's in the crate), give her a quick peck, and then immediately remove the treat container from the crate. As you are doing this, immediately click and treat her again. It's helpful that she's in the crate and that there is only one exit because you can use your hand with the treat container as a body block to keep her from walking out of the crate. Do the entire process five or six times very quickly: click, treat, remove treat container from crate, immediately click and treat again.

After you've done this five or six times, and you've been able to "catch" your chicken with treats before she walks out of the crate, allow her the opportunity to decide if she is going to stay in the crate or walk out of it. Hold the treat container to your stomach, covering it with your other hand, and stay still for just a moment—*just a moment*, though. Count to one or two and then, if she's still in the crate, click and treat again. If she walks out of the crate, use your target training to lead her back inside and start again.

Your chicken will quickly learn that when she's in the crate, it's awesome because she gets loads of yummy treats. When she's out of the crate, nothing particularly exciting happens.

Work on this behavior for only a few minutes and then give your chicken a break (preferably in a different crate—one for training, one for hanging out). I should note here that I have two crates in my house for my chickens. One is large enough to walk around in, with shavings on the bottom, a box for laying eggs, and food and water containers. Because I bring a chicken indoors for a few days for training, this is her home away from home during training. My training crate is the one I use for transport and is a little larger than a cat crate. There's enough room for her to stand, turn around, and lie down comfortably, and there's a towel in the bottom, but it's a little more snug than the "chicken hotel room." Because this crate is used only for transport and training, a chicken is only in it for short amounts of time.

After you've done this a few times, and your chicken is willingly walking into the crate in a fairly fluid motion, you don't have to click and treat as often. Start clicking one or two times during the entire sequence of turning her head, walking toward the crate, walking into the crate,

ADAPTING TO THE CRATE

Surprisingly, chickens adapt fairly quickly to crates. I think it's because they enjoy the safe, snug feeling of a nesting box. They are covered from the top and sides, so they don't have to worry about predators. All of my chickens (roosters included!) become calm within minutes of being crated. The training part of crating comes into play when a chicken is in the crate longer than she would expect (she is usually ready to move around after a couple of hours). She needs to get used to being in a different environment with different sounds, smells, and even temperatures (in your house or garage versus the barn or coop). She needs to learn to walk willingly into the crate versus being placed in the crate with the door quickly closing behind her, before she can escape. Being carried in a crate, having her crate in a moving car, and even just seeing people or other animals dogs come close to the crate door may be worrisome for a chicken before she is accustomed to it. Simply being in the crate is the easy part.

turning around in the crate, and staying in the crate. Remember, though: in the beginning, you're clicking and treating with every single step.

If your chicken is having a hard time, offer a jackpot for each milestone. Milestones for this behavior are:

- Turning toward the crate
- Walking through the threshold of the crate door
- Turning around in the crate
- Standing still in the crate

When you've made it through all of these steps, you may even find that your chicken lies down in the crate—with the door open! Chickens get bored pretty easily, so she is either going to want action, or she will enjoy the peace of the crate. If she wants action, simply work on starting to delay the rate of reinforcement—that is, extending the time between the click and the treat. You can create a longer and longer duration and get to the point where you can even click two or three times before offering the treat (that's advanced chicken training!).

When your chicken is doing really well with all of the steps, you can start to wean her off the target stick and treats. Don't wean her off the treats too quickly, though, or else you'll end up with a bird that is no longer interested in working for you.

WEANING OFF THE TARGET STICK

Up until now, you've been holding the target stick (wooden spoon) in your hand, and it's probably reaching out 6 inches or more. When you start to wean your chicken off the target, you'll eventually want her to follow just your hand or a pointed finger (and not necessarily peck at it the whole time!). If you're holding the target a couple of inches away from her beak, and she is walking toward it, you can probably get away with her not actually pecking at it. If you are too slow, she will catch up and peck at it. However, if you're too fast, you may move the stick away from your chicken too quickly, and she will have to run toward it. If the stick gets too far away from her, she'll lose interest and move on to something different. Be sure to match your pace to your chicken's. When my chicken fully understands what the target stick is, I will move it fast enough so she doesn't actually peck at it the whole time.

When you want to start weaning your bird from the stick, just gradually hold more of it in your hand so that less of it is sticking out. Initially, there may be 6 inches of the stick protruding from your hand, then 5 inches during the next session, and then 4 inches during the session after that, and so on until your finger sticks out farther than the target stick. Eventually, your finger is all that your chicken sees and follows. (I like to use two fingers—no real reason other than it's what I do with my dogs, so I use the same visual with my chickens.)

Once you have your chicken following your finger, you can start weaning her off the targeting during crate training. If you start off with her following the target all the way into the crate and then luring her to turn around in the crate, take baby steps to get to the point at which she will willingly go into the crate, turn around, and lie down when you simply point your finger at the crate—with only one click and treat at the very end!

When your chicken is following your finger, make it clear to her that you are pointing at the crate. If she is looking the other way, move your finger near her face so she can clearly see you. Then, direct your finger toward the crate. You may have to go backward in your training and click-treat for more behaviors than you were when you were using the target stick. Sometimes, your chicken may not generalize, and she may think that she is learning a brand-new trick because the target stick is no longer there.

Target (finger point) to get your chicken to go to the threshold of the crate. You can even move your hand into the crate, but don't lure your chicken to turn around—just wait and see if she does it on her own. She may back out of the crate, or she may turn around and stop—jackpot! Your chicken just walked into the crate on her own and stopped. You've also just "chained" your chicken's behaviors, which means that you had her do multiple behaviors in a row before treating her.

Tell your chicken "OK" and stop treating her. Wait for her to walk out of the crate, and then do the chain of behaviors again. Each time, make your target (finger) a little bit less noticeable than the previous time. Example: If you had to move your hand halfway into the crate to get your chicken to go all the way in, then next time move your hand only a quarter of the way into the crate. Then, when your chicken is doing well with that, move your hand only an eighth of the way in. Eventually, your hand will be just at the door of the crate, pointing.

Intermittent reinforcement is a good practice to use with crate training. Once your chicken is doing really well with it, you shouldn't stop reinforcing her completely. I suggest starting to reinforce her (click-treat) based on her performance of the behavior. Say your chicken goes into the crate, but she decides to look around before she does. You might still give her a click and a couple of pecks. But, if you pointed at the crate and she went directly into it and promptly lay down, I'd give her a jackpot!

Chickens have many different vocalizations, so you can put this behavior (and other behaviors) to a verbal cue, and your chicken would learn to understand it. Start by saying "kennel" while you point to the crate. Keep using the verbal cue as you are weaning off the target stick/pointing, and your chicken eventually will respond to just the verbal cue. Give it a try and see what happens!

Trick: Jumping onto Your Arm

Before we start, please note that heavy-boned chickens and older birds may not be the best at this trick. Consider your chicken's limitations when working with her.

I think it's a chicken's natural instinct to move away from something that is moving toward her, so when you want to teach your chicken to come to you, do not move toward her. In the beginning of this trick, lower your arm and *then* start to ask your chicken to come to you. If you ask your chicken to come to you and then lower your arm, she may just walk away. Try to make your movements slow, but with purpose. In other words, go slowly, but not in a creepy way, or else your chicken may think that you're up to something and not want to come near you.

I start this trick on the table. I love to work with my girls on the table because they already recognize this as an area for training (and getting treats!), so they are more likely to stay on the table and engage with me. You still want to work on baby steps, so once your chicken is doing well with this trick on the table, you work on the floor. When you place your chicken on the floor, go down on the floor with her. This way, you're neither asking her to jump up to a higher place nor leaning down into her space, which may be scary for her in the beginning.

Perching is something that our chickens do naturally, but perching onto something that may be wobbly and unstable isn't something they enjoy. I suggest doing this trick at first on your bare arm—no sleeves. Your sleeve may move around on your arm, making it unstable, and then your chicken will probably want to hop off. Just to warn you, though, she may grab on with her nails, so you might get a little scratched up as the two of you are figuring out this trick.

Here are the steps to get started:

- Have your chicken on the table.
- Hold your treat container and clicker.
- Place your arm, bent in an "L" position, toward your chicken.
- Hold your treat container on the opposite side of your arm from your chicken so that your chicken will be moving toward your body to hop onto your arm. Note: If your treat container is too close for your chicken, she may simply lean forward and be able to reach the treats. If your treat container is too far from her, she'll probably just hop over your arm to get the treats. You'll have to play around with this a bit to find a happy medium.
- Hold your bent arm so that it is hovering over the table, just high enough so that she'll have to step up to get on your arm.
- As soon as she puts one foot on your arm, click-treat. The first couple of times, she may step up with one foot while keeping her other foot on the table. That's OK! Click and reward—remember, baby steps.

Once she's comfortable with having one leg on your arm and getting rewarded, slowly lift your arm up just an inch higher when she puts one leg on your arm. This will make her have to shift her weight, and she'll either step down (hopefully not), or she'll lift her other foot onto your arm—*ta-da*! She's now perched on your arm. Give her a jackpot!

If your chicken enjoys being petted, this would be a great time to just stand calmly and pet her. Or, she may be in "working mode" and not really want to be petted at the moment. She may just want treats, and that's OK, too.

Once she's comfortable stepping onto your arm, you can move your arm a little higher off the training table.

Once you have her stepping onto your arm nicely while she's on the table, work on proofing her by doing the same action in different contexts, such as on the floor. Kneel or sit on the floor the first few times so your chicken can still step up, rather than having to leap, onto your arm. You can place the crate on the floor and try having her step up onto your arm as soon as she walks out of her crate. You can also start to put a cue, such as "up" or a kissing noise, to the behavior. Sometimes, even just the sound of the clicker and the visual of your arm is enough to make your chicken realize what she needs to do to get the treat.

Remember—operant conditioning tells us that if we keep rewarding a behavior, the behavior will never progress. So, if your end goal is that your chicken actually jumps into the air and lands on your arm, you should not continue to reward her for simply stepping onto your arm. Once your chicken is pretty consistent with stepping up, stop rewarding that behavior and start raising your arm higher and higher, just a few inches at a time, so that she has to start to jump for it. Don't make it too hard for her, or else she may give up. Keep her confidence up. If she jumps for your arm and misses, go backward in your training for a few sessions to help rebuild her confidence. In no time, you'll be able to walk near your chicken and give your command, and she will jump up onto your arm.

Chapter 7

Chickens, the Environment, and Other Animals

H aving chickens can be wildly fulfilling. I love to sit outside and watch my chickens roam about. The little sounds they make are adorable. It's comical how if one chicken finds something to eat, she will quickly run away with it, ultimately alerting the other chickens that she's just found something tasty. Ironically, if she would have just eaten it when she found it, she wouldn't risk having all of the other chickens running after her to steal her treasure. Goofballs!

I also find it interesting to study the hierarchy of the flock. My friend, Heather Lockhart, believes that the alpha hen is the first one in the morning to lay her egg. After that, the others will lay. I haven't been in my barn for extended hours to notice, but I bet each chicken lays her egg around the same time. I do notice that some are late-morning layers and some are afternoon layers.

There is so much to learn about chickens! We know that these adorable creatures have social hierarchies, have many different sounds for communication, and are masters at body language.

(Yes, I can tell when our rooster, Marlboro, is in a mood and about to go after me!) Did you know that, when you give chickens the right lifestyle, they're also sustainable and useful for the environment—and beneficial to other animals? Yep, our happy, silly little chickens!

Many of my horse friends swear that as long as they have horses, they will have chickens. Chickens are wonderful for horse pastures. They

Training the chickens' buddies (from left to right) Daisy, baby Petunia, and Lily.

will kick up the manure and spread it around, gobbling up bugs and worms while they do it. They eat insects and worms that could potentially get into the horses' digestive tracts, while keeping the pastures healthy. It's a win-win!

I have potbellied pigs living in the barn with my chickens. I don't think that the chickens are necessarily beneficial to the pigs, but they do get along great. The pigs make nests with their straw each night, and the chickens love to dig around in them and look for bugs. When the weather is nice, the pigs will spend a good part of their day rooting around, looking for grubs, and roots, and there is always a chicken nearby to gobble up what the pigs missed.

You don't need to do much training to help horses and pigs get along with chickens. However, many of us who have chickens also have dogs. Or, as our chickens see them, wolves and coyotes. Now, most of the time, the family dogs are pretty good with the chickens. They don't usually bother the chickens unless a chicken takes off running and sets off the dogs' prey drive. Some dogs have a higher prey drive than others. Many dogs are content to just hang around and let the chickens be chickens, but some of our dogs need a little extra management.

If you have a dog that wants to chase (and possibly eat!) your chickens, there are a few things you can do to help remedy the problem. Just like we use shaping to train our chickens, we can shape our dogs to understand that it is to their benefit to *not* chase the chickens. If you have a dog with a high prey drive, you may want to skip right to the Management sidebar in this chapter—just like with many animals, you can't train out instinct. But, if you have a dog that thinks it's fun to chase the funny thing that runs away, there are some steps you can take.

Use what you've learned in this book and shape your dog to stop chasing the chickens. Start off with a controlled situation, a bag of high-value treats, and your clicker. (You will have already conditioned your dog to understand what the click means.)

Have a friend hold a chicken or, better yet, have your chicken in a crate. Have your dog on leash with you. You are going to train your dog that whenever he sees the chicken, he looks at you. It's pretty simple but highly effective. Allow your dog to understand the concept of "in the presence of a chicken, I look to my owner." Think of it this way: If your dog wants to chase your chicken and is fixated on the chicken, you have little hope of getting his attention. But, if your dog wants to chase your chicken but looks at you before he does so—*bam!* You now have your dog's attention, and you can direct him to do something else.

With your dog on leash and your chicken in a crate, click and treat your dog each time he looks at the chicken. Dog looks at chicken; you click and treat. Sounds a little counterproductive, eh? *Why would I want to reinforce my dog for looking at the chicken*? I'm glad you asked!

TIME TO MOVE ON

Remember how we say that once an animal is doing a behavior about 80 percent of of the time, it's time to move on to the next step? If you're thinking, *I don't even know if my dog actually realizes that there is a chicken in there*, that's when you know that you're ready to move on to the next step.

First off, you're not rewarding your dog for chasing the chicken. You're certainly not going to click and treat if your dog is pawing at the cage or barking at the chicken. You are starting off by rewarding your dog for calm behavior and for simply being near the chicken.

Second, think about how the clicker works: it's a marker, and you're marking the behavior of looking at the chicken. If your dog has been conditioned to understand that the click means a treat is coming, theoretically he should hear the click and then immediately look up at you in anticipation of the treat, right? So he looks at the chicken, gets clicked, immediately looks up at you, and gets a treat. *Hmmm…*

Dog looks at chicken. Click-treat. You should start to get quick back-and-forth action from your dog. He'll start to very quickly move his head toward the chicken but then will look back up at you as fast as he can. You may even find that he will start to just kind of turn his head but, after

a few times, won't even really be looking at the chicken. That's fantastic, and a great time to give your dog a jackpot. He knows the chicken is there, but he is choosing not to look at her because he already knows the next step.

When your dog is bypassing actually looking at the chicken, you can start to make the exercise harder. For example, you can take your dog and stand on the other side of the crated chicken with him. Or put the chicken (still in the crate) on a chair or table so you are changing the height of the chicken. With good training and timing, your dog should be thinking, *Yeah, yeah, yeah, I see the chicken, but just give me the treat*!

Keep changing the situation by moving the chicken around more, working inside and outside the house, and changing the chicken's height again. Try having a friend hold the chicken versus having the chicken in a crate (or vice versa, depending on how you started). Put the dog on a longer leash; this allows him the freedom to make more of a decision about what to do.

When you have proofed your dog with the chicken in many different ways, you can think about allowing your chicken to run free in the yard. Dogs are still animals, so keep him on leash until

Observing your chickens can teach you a lot about individual and flock behavior.

MANAGEMENT

So you have a dog that thinks your chickens are yummy meals in the yard just for him? Well, training will certainly help, but I wouldn't trust him. If you know that your dog will kill a chicken, it's your responsibility to keep your chickens safe, and do not fault your dog for doing something that simply comes naturally to him. Invest in some good fencing. Have a nice, sturdy tie-out for your dog. Talk to everyone who lives with you and leave notes on the door so no one forgets and accidentally has the chickens and dogs out at the same time. You will have to manage the situation to make sure everyone stays safe.

you feel extremely comfortable with his training. At this point, a moving chicken shouldn't faze him because you will have proofed him with enough chicken distractions that he should pretty much ignore the chicken altogether. He will be conditioned to know that, in the presence of a chicken, he pays attention to his owner (and gets rewarded for it!). I love shaping!

Chapter 8

Let's Talk about Roosters

8

used to say, "Life's too short to have a mean rooster." I still believe that, but not as much.

Our first rooster was Frodo, an enormous black Cochin. My friend ordered chickens from a hatchery and received a "bonus chick" with her order. (I've learned so much about chickens since then, and, looking back now, I know to certainly be wary of those bonus chicks! It's probably just a way for hatcheries to get rid of male chicks.) So, I went to my friend's house to see the chicks and pick some out to bring home. We picked five little chicks of different colors and were enamored by the little bonus chick. This little peeper would come right up to my hand and peck at my shiny ring. How adorable, right? Ha! Now I know to watch out for the chick that has the guts to come right up to you and start pecking. I'm not saying that they all turn out to be roosters, but it's a pretty good indication.

Another interesting thing about our bonus chick was that the other chicks seemed to despise this little guy. They would peck at him any chance they could get. Now I think that, yep, they knew he was trouble from the get-go!

Little Frodo was adorable until around four months of age, when he got his crow. Once he had his voice, he was the man! He started rushing us when we'd come into the barn, so forcefully that we would have to enter the barn with canes in our hands. I love animals, so when I say that we were walking around with big sticks, we were not actually hurting our rooster with them. The sticks were purely for self-defense. Just holding a stick out toward him seemed to be enough to keep him at bay; otherwise, he would rush you and fly up at you and try to get you with his spurs.

It got to the point where he would see us the moment we came out of our house and he would literally run up to us at full speed. And did I mention that Frodo grew to be the biggest chicken I had ever seen? I swear, he almost came up to my waist. Now, I've worked with 120-pound Rottweilers and German Shepherds, so I don't want to sound like I was afraid of this chicken, but… I was afraid of this chicken! He was fast, and he was mean. He thought of three things in life: eating, mating, and making our lives miserable. If I didn't have three kids, I might have considered keeping him. *Might have*.

My youngest child at the time was four years old. Frodo was taller than he was. It really did come to a matter of safety. Luckily, we had a friend whose father lived on a farm. He had horses and potbellied pigs and ducks and all kinds of chickens. He was a big, gruff older gentleman—the type that you'd assume would take in a rooster just to have something yummy in the slow cooker that night. But no—he was a total animal lover! We asked him if he'd take Frodo, and he said that he would.

Catching Frodo and putting him in a crate to take him to the next town was not an easy task. I searched my home for what I could use as a makeshift full-body suit of armor. I don't fully recall, but I think a ski

Frodo—before he was the same size as Corey!

mask was part of my ensemble, and I definitely wore big, thick gloves. I managed to wrangle Frodo into a crate and poured myself a cold glass of wine afterward. I brought his crate to our patio and sat down next to him to drink my celebratory wine. As I talked to him and told him how handsome he was (although he was a total jerk, he truly was a beautiful chicken), I put my fingers near the outside of the crate door. I don't know how he did it—it all happened so fast—but he managed to get my hand from the inside of the crate. Ouch! He had the back of my hand in his beak and wouldn't let go. Little turd… All he was doing was reinforcing my decision to rehome him.

That evening, we took Frodo to his new home. We warned his new owner that, although he was beautiful, we did consider him a dangerous rooster. Very patiently and quietly, the man looked at Frodo through the crate. They eyed each other up. He slowly opened the crate door

My friend Jill with a sweet rooster named Sonny at Heartland Farm Sanctuary.

(while my son moved to stand behind me) and reached in. He brought Frodo out and gently tucked the rooster under his arm. The two of them seemed frozen in time, just looking at each other. He said a few quiet words to Frodo and then gently placed him on the floor. Frodo stood there for a moment and then quietly walked off to meet his new flock of hens.

We were flabbergasted to say the least. It's like the two of them had this unspoken understanding. Ever since then, I have referred to Frodo's new owner as the "rooster whisperer."

We've had a few other roosters in our lives—some jerks, some actually pretty sweet. All of them have been wonderful to their hens. I know that when roosters try to attack people, they are really just doing the job that has been encoded in them for the past million years. I don't dislike roosters now, but, if I'm being honest, there was a time when I really didn't like them. I saw them as little jerks that assaulted the hens to no end and tried to spur us. Now that I know better, I really do have an appreciation for them. First off, even the plainest rooster is truly

CAN YOU TRAIN A ROOSTER?

I don't see why not! Roosters can be just as smart as hens, right? However, once their testosterone kicks in, they certainly have other things on their minds. They are smart creatures, but you cannot train out instinct. If you have a rooster that you want to train, I suggest starting young, before adolescence hits at around four months of age. If you have only a rooster, and no other chickens, you're probably pretty safe. He might decide that you are his hen and become possessive over you, so be wary of that. But, for the most part, single roosters can be pretty good pals.

If you want to train your rooster, but you also have hens, be sure to bring your rooster away from his girls for training. Crate-train him and keep him far enough from them so he can't hear, smell, or see them. If he can, all bets are off! He'll have a one-track mind and will want to get back to them. You may have to use a little more patience with a rooster than with a hen, but I'm sure you'll have just as much fun training him.

beautiful. Second, roosters really do take care of their girls, showing them where the food is, herding them under bushes if they see hawks, letting them eat first, and offering up their lives to protect them against predators. They really are the "front line" if a raccoon or other predator were to get into the barn. Their instinctual rooster brains often don't see the difference between people and predators coming in to hurt their flock. It also helps that our roosters don't attack me as often as they do my husband!

Our main man right now is Marlboro. He's a handsome fella that dotes on his girls with undivided devotion. He got his name because when he first started to crow, he sounded like a little old lady that had been smoking cigarettes for eighty years. Marlboro rarely goes after me, but he'll go after my husband and my sons. I try to tell them that if they just move toward him slowly and put their hands out as if reaching toward him, he'll back up and leave them alone. If someone enters his space calmly, he will be respectful. So either they're not trying, or it's simply not working for them. My youngest has to go into the barn with a rake in his hand for protection!

As I said, life's too short for mean roosters. But I know my girls are better off with them around.

Chapter 9

Chicken Enrichment

I love enrichment ideas for my chickens, not just because chickens are smart and inquisitive, and I enjoy doing things that make them happy, but also because I want them to really be able to use their minds and think about things. The more you allow your animals to think and discover, the happier and more well-rounded they will be. I see chickens that are stressed, or in close quarters with other chickens, or simply bored. They don't have anything to *do*! These little birds are smart and active, and they want to do things that are fun and challenging. Just like little kids and dogs, if you don't give them something to do, they are going to find something to do. In the case of chickens, this will likely include digging holes and pecking at other chickens. You'll enhance their quality of life by providing them psychological and physiological well-being. In other words, give them something to do, and they'll be happy chickens. In addition to training (and lots of it!), here are some ideas for enrichment activities to do with your birds. While some of these suggestions work best for outdoor chickens, many work just as well for indoor chickens.

Puzzles: There are a ton of puzzles marketed for dogs, and I've also used them with my chickens. I did a little behavior experiment at home with some of my animals: a dog, a pig, and a chicken. I wanted to gauge intelligence by putting their favorite treats in the puzzle and then seeing who could get their treats out the fastest. Can you guess who won?

Well, I don't know how realistic the test was. The dog tried to pick up the puzzle. (He's not my brightest pup, but he certainly is the sweetest!) In his defense, the puzzle was made out of wood, and he does like to carry wooden sticks around. So he wasn't very quick at getting to the treats, but he did eventually get the puzzle open.

Puzzles made for dogs can be fun for chickens, especially if there is food involved.

The pig was surely the most determined. Pigs love to eat just as much as chickens do! But I think the pig had a limitation because she was not truly able to paw at the puzzle. She had to use just her snout to get the puzzle open. She received our second-place ribbon.

Finally, we presented the chicken (Sam, my Andalusian) with the puzzle. *A-ma-zing*. The puzzle had four "pockets" that the animal had to slide open to get to the treats. Sam was used to me presenting her with items meant for problem-solving, so I knew she would be curious. It took her no time at all! She looked at the puzzle and immediately started to scratch at it. *Bam*! She got the first pocket open. She looked at the second pocket, stood up, and scratched it. *Bam*! The second pocket was open. I had a feeling that she would be pretty amazing at this puzzle, but I love it when I'm shocked at the intelligence of my animals.

Hanging/buried snacks: Place snacks on a shelf, dangling down so your chicken knows that they're there and has to hop up in the air to reach them. Hang snacks, such as lettuce leaves, to strings attached to clothespins. Hanging watermelon slices are perfect—the chickens will have to reach up and around to get at them, and they won't get messy from laying on the coop floor. Day-old bagels are also perfect for hanging because they have handy premade holes. I also like to take dog biscuits and hide them in the holes that my pigs dig in their pen; the pigs and the chickens both get in on this treasure hunt!

Sam is one of my all-time favorite chickens.

Meaty bones: Yep, chickens are little dinosaurs! My dogs love meaty bones, and so do my chickens. The chickens try to drag them away from each other in games of keep-away.

Snack balls: You've probably seen balls for dogs and cats that have the holes in them. There is the hard plastic kind (that often has a smaller ball inside), and the rubber kind, which looks like a human baby toy. You can buy something similar for your chickens, or you can make one using a wiffle ball or some cardboard. Put lettuce inside so the chickens have to peck at the holes to get it out. You can also fill a hard rubber dog toy (such as a Kong) with treats. Use a little peanut butter to hold everything in place. Or put frozen peas inside the toy so that each time the chickens peck at it, a few peas roll out.

Snacks on a string: Cheerios? Popcorn? There are many foods that can easily be strung together for your chickens.

Snacks on a log: Spread oatmeal, cooked rice, or pieces of cooked pasta on the log for the chickens to scrape off. It is a great opportunity for them to use their minds because they'll have to get the food out of all of the little nooks and crannies.

Something new: Provide your chickens with new, different foods couple of times each week. Our chickens will eat just about anything, and they are thrilled when they get something new!

Chicken scratch: A block of chicken scratch from the pet-supply store provides the chickens with something to do on a winter day when there are no bugs to find.

Ice wreath: A ring of ice with treats frozen inside is great on a warm summer day.

Flakes of hay: I toss a flake of hay in for my chickens and pigs almost daily. Pigs need it to munch on, and they like to make their nests out of hay and straw. The chickens love to hop on top of it, scratch it, and spread it out to see if they can find any bugs.

Treat-dispensing toys provide mental exercise and tasty rewards.

Coop swings: I don't currently have one, but I've seen these cute little swings that people put in their chicken coops. The chickens seem to enjoy sitting on them, and it's pretty adorable to watch. I wonder if they'd enjoy a little hanging basket or hammock, too.

Snuffle mat: These are made for dogs but would be easy to make for chickens, too. Hide a few mealworms in the mat to stimulate your chickens' scratching instinct.

Hiding spots: Hang up the bottom of an unused, old-fashioned mop so the chickens can go underneath it to play hide-and-seek.

Plastic bottle feeder: Poke a bunch of holes in an old plastic water bottle and then fill it with your chicken's food. Make sure that there are plenty of holes and that they are large enough for the food to come out. Let your chickens peck and scratch at the water bottle; when it rolls around on the ground, the food will fall out.

Cricket tubes: Fill a tube with crickets or other edible bugs, and make sure that the holes are big enough for the bugs to fall out. The chickens will love it!

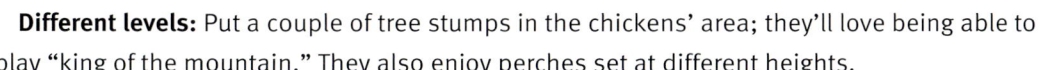

Different levels: Put a couple of tree stumps in the chickens' area; they'll love being able to play "king of the mountain." They also enjoy perches set at different heights.

Sunflower: Hang a sunflower in the coop or, better yet, let one grow in there. Sunflowers offer great treats and enrichment. The chickens have to jump up to peck at the sunflower, and then yummy sunflower seeds rain down.

Sandbox: My girlfriend took an old tire and filled the inside with sand, so her chickens have their own little sandbox for dust baths.

Obstacle course: Create a small obstacle course that your chicken has to navigate in order to get her food. Start with something simple, with her food easily in view, and keep adding to it.

If you have indoor chickens you may want to provide an enriching environment for them. Add fake plants for them to hide under. Use a puzzle as a feeder, so they have to work for their food, not just get it easily from a container (that's no

Appendix:
Chicken Rescue

On a certain level, I feel sad for all of the animals that humans have decided are useful. Once humans decide an animal is useful to them, the quality of life for that species often becomes intolerable. Chickens certainly are near the top of this list. Humans eat chickens. Americans alone consume more than eight billion chickens a year. We eat the eggs that they lay. We use their feathers to make everything from shoes to dishes to circuit boards to diapers. Male chicks are deemed undesirable and are either ground up alive or are kept in dark basements as fighting cocks, with their wattles and crowns cut off with scissors. Hens are selectively bred to produce eggs on a daily basis, which often results in serious and traumatic body failure because they are laying about 350 percent more eggs than they were meant to lay. When they can no longer lay daily, they are turned into pet food. Chickens are so amazingly adaptable that they can survive in horrendous factory-farming conditions that seem almost unlivable. There is a very dark side to the lives that billions of these birds are born into.

Chickens are God's creatures, and they deserve to be treated with respect. They live and breathe. They have personalities. They can be afraid. They can become old and weak. They have close friends. They feel pain. They get cold and are susceptible to frostbite. If you have heard a mother chicken calling to her youngsters, there is no doubt that you can see her devotion and joy. These poor creatures have endured centuries of human torture, and they don't deserve it.

Scientists have uncovered that chickens are the closest living relatives of the mighty Tyrannosaurus rex, the most feared and famous of all dinosaurs. That alone should command some respect! My friend Susan, former owner of a chicken-supply store, once said, "I'm pretty sure chickens would eat us if they could." Seeing as they are basically miniature T. rexes, I believe her!

My chickens are pets. I am responsible for them and their well-being. I love it that they lay beautiful, delicious eggs, but I consider that to simply be a bonus. They live their lives with us, even when they stop laying. Their home is in our spacious barn with organic feed and scratch. I take them to the vet when they don't feel well. They are allowed to lie in the sunshine when they feel like, it, graze and search for bugs, and take daily dust baths. They have pretty good chicken lives.

Many of my friends who keep chickens provide them with similar lives. They care for and love their chickens. I think that the number of people who keep chickens as pets is going to continue to grow. If someone would have told you five hundred years ago that he or she was going to keep a dog for no particular purpose and would allow the dog to sleep in bed with him or her, you'd probably think that this person was bonkers! Well, it's same thing with chickens. Keeping chickens may seem strange to some, but, in time, more and more people will think that it's cool to keep chickens as pets.

So, it's a sad time for chickens, but it's also a pretty amazing time for them. Chicken rescue groups and sanctuaries have been popping up around the globe. Volunteer animal lovers are devoting their time and resources to help these regal beauties find solace in the second part of their lives. If you can, please help support these sanctuaries. Most of them are doing a wonderful job to help the birds and other animals. You can often take tours and learn about what they do. If you are thinking of getting chickens, volunteering at a sanctuary is a great idea. You'll get to spend time with the chickens, and I'm sure you'll learn a lot.

There are some pretty amazing rescues and sanctuaries in my area. Special mention goes to:

Farm Bird Sanctuary
www.facebook.com/FarmBirdSanctuary

Heartland Farm Sanctuary
https://heartlandfarmsanctuary.org

Nutzy Mutz and Crazy Catz
http://nutzymutz.com

Index

About the Author

Giene Keyes started working with animal behavior in 1990 when she adopted a Greyhound from the racing track. To date, he is the most severe separation anxiety case that she has seen, and he taught Giene about the psyche and learning process of dogs. In 1995, when she was training her little potbellied Labrador Retriever puppy, she really became interested in discovering how dogs learn and how best to train them. Starting her small training business in 1997, she developed her own methods and ideas of positive reinforcement. By taking what she had learned from her Greyhound and from using positive-training techniques with horses, she created what is now the foundation of her training methods with chickens and other species.

As a six-time *Madison Magazine* Best of Madison Gold winner and *Pet Age Magazine* National Best Business Award Gold winner, Giene is a professional dog trainer and behavior specialist in southern Wisconsin. Starting one of the first dog daycares in the Midwest, she is a pioneer in her field. She devotes her time to animal behavior and enhancing the lives of people and their pets. Specializing in aggression and other difficult dog cases, Giene also works with rescue groups and shelters, evaluating dogs for adoption, developing canine aggression management programs, and educating staff on dog language and pack mentality. She has provided behavior consultations for service dog organizations, rescue groups, and dog daycares throughout the Midwest. It wasn't until she served on the board of directors for the Heartland Farm Sanctuary that she discovered a passion for pigs and chickens.

A seasoned public speaker, Giene routinely presents seminars on animal behavior, training, and operant conditioning. People flock to her chicken training workshops, where they learn the principles of team building in a fun and interactive environment. Giene is a regular guest on popular radio shows, newscasts, and podcasts, and she has been featured in local magazines: the *Wisconsin State Journal* highlighted her work with chicken clicker training, and she graced the cover of *Wisconsin Woman* in 2010. She continues to love sharing her knowledge about and the joys of training chickens.

When she is not working, you can find Giene at her farmette in the rolling hills of southern Wisconsin, spending time with her family and her dogs, cats, pigs, and, of course, her clever chickens.

Website: *www.gienekeyes.com*
Facebook: *Facebook.com/ChickenClickerTraining*
Instagram: @chicken.clicker.training

Photo Credits

Front cover: chicken, stockphoto mania/Shutterstock; clicker, DenisNata/Shutterstock

Contents page/sidebar chicken and footprint graphics, back cover, pages 3 – 5, 19, 20, 30, 34, 35, 43, 61, 65 – 67, 81, 85, 86, 92, 98, 105, 106, 111, 113, 121, 129, 135: katyalitvin/Shutterstock

Cluck the Chicken Store: 31 (bottom), 35

Tina Gerber, 14

Courtesy Giene Keyes: 7, 9 – 11, 16 (right), 18, 20, 22, 23, 28, 30 (top and bottom), 31 (bottom), 36, 37 (top), 38, 40, 42, 43, 46, 47 (top), 49, 50, 52, 62 (bottom left and right), 63, 69, 85, 87 (top), 91, 100 (left and right), 103, 118, 119, 122, 127, 128, 134, 142

Abby Wicker: 16 (left), 24, 25, 47 (bottom), 48, 58, 60 – 62 (62 top), 64, 67, 71, 88, 96, 101 (top, center, and bottom), 104, 107, 108, 112, 115

The following images courtesy Shutterstock:
alexei_tm, 133; Big Foot Productions, 126; WilleeCole Photography, 21; cynoclub, 57, 120 – 121; Tracey Helmboldt, 117; Eric Isselee, 19 (left), 123 (bottom); Kpuleo, 123 (top); Dmitriy Krasko, 37 (bottom); Erik Lam, 19 (right); l i g h t p o e t, 83; Louella938, 59 (bottom); Oleksandr Lytvynenko, 137; Anastasiia Malinich, 135 (bottom); B-D-S Piotr Marcinski, 68; Milosz_G, 45; monticello, 55; Nach-Noth, 132; DenisNata, 56; Aksenova Natalya, 97; Olhastock, 1; Lena Pan, 70; pets in frames, 27; ANURAK PONGPATIMET, 86; PRESSLAB, 125; pryzmat, 134 (bottom); Suthida Sririttha, 111; Kuttelvaserova Stuchelova, 135 (top); Galyna Syngaievska, 51; Tsekhmister, 8, 84, 89; TTstudio, 13, 95, 131; Twin Sails, 33; Wild As Light, 41, 90; yevgeniy, 39; yevgeniy11, 93; Ratmanant Yotsurin, 59 (top)